D1716767

COMPACT Research

Disaster Response

by David Robson

Current Issues

San Diego, CA

For more information, contact:
ReferencePoint Press, Inc.
PO Box 27779
San Diego, CA 92198
www. ReferencePointPress.com

Picture credits:
Photoshot: 11, 15
Steve Zmina: 31–34, 46–48, 60–63, 75–77

LIBRARY OF CONGRESS CATALOGING-IN-PUBLICATION DATA

Robson, David, 1966–
Disaster response / by David Robson.
p. cm. — (Compact research series)
Includes bibliographical references and index.
ISBN-13: 978-1-60152-081-4 (hardback)
ISBN-10: 1-60152-081-6 (hardback)
1. Emergency management. 2. Emergency management—United States. 3. Disaster relief.
4. Disaster relief—United States. I. Title.
HV551.2.R63 2009
363.34'80973—dc22

2009002283

Contents

Foreword

“Where is the knowledge we have lost in information?”

—T.S. Eliot, “The Rock.”

As modern civilization continues to evolve, its ability to create, store, distribute, and access information expands exponentially. The explosion of information from all media continues to increase at a phenomenal rate. By 2020 some experts predict the worldwide information base will double every 73 days. While access to diverse sources of information and perspectives is paramount to any democratic society, information alone cannot help people gain knowledge and understanding. Information must be organized and presented clearly and succinctly in order to be understood. The challenge in the digital age becomes not the creation of information, but how best to sort, organize, enhance, and present information.

ReferencePoint Press developed the *Compact Research* series with this challenge of the information age in mind. More than any other subject area today, researching current issues can yield vast, diverse, and unqualified information that can be intimidating and overwhelming for even the most advanced and motivated researcher. The *Compact Research* series offers a compact, relevant, intelligent, and conveniently organized collection of information covering a variety of current topics ranging from illegal immigration and deforestation to diseases such as anorexia and meningitis.

The series focuses on three types of information: objective single-author narratives, opinion-based primary source quotations, and facts and statistics. The clearly written objective narratives provide context and reliable background information. Primary source quotes are carefully selected and cited, exposing the reader to differing points of view. And facts and statistics sections aid the reader in evaluating perspectives. Presenting these key types of information creates a richer, more balanced learning experience.

For better understanding and convenience, the series enhances information by organizing it into narrower topics and adding design features that make it easy for a reader to identify desired content. For example, in *Compact Research: Illegal Immigration*, a chapter covering the economic impact of illegal immigration has an objective narrative explaining the various ways the economy is impacted, a balanced section of numerous primary source quotes on the topic, followed by facts and full-color illustrations to encourage evaluation of contrasting perspectives.

The ancient Roman philosopher Lucius Annaeus Seneca wrote, "It is quality rather than quantity that matters." More than just a collection of content, the *Compact Research* series is simply committed to creating, finding, organizing, and presenting the most relevant and appropriate amount of information on a current topic in a user-friendly style that invites, intrigues, and fosters understanding.

Disaster Response at a Glance

Millions of Suffering People

Natural disasters such as earthquakes, wildfires, flooding, and volcanic eruptions killed 220,000 people in 2008. Man-made crises, including war, terrorism, and plane crashes also took a catastrophic toll. Millions were killed, injured, lost homes, or lost jobs.

The United Nations

The United Nations (UN)—one of the primary relief agencies—serves some 33 million refugees and displaced people in over 110 countries around the world.

Nongovernmental Organizations

Organizations such as the International Red Cross, Doctors Without Borders, and Oxfam play a key role in serving victims in dangerous and remote locations such as Sudan and Burma.

Distributing Resources

Relief organizations distribute approximately 17,000 tons (15,422 metric tons) of food to 891 million victims of disasters throughout the world each year.

Coordinating Responses to Major Disasters

During a major disaster, a coordinated response is essential but often difficult to accomplish because so many different governments and private organizations take part.

Paying for Disaster Relief

While some of the funding for disaster response and relief comes from private donations, most of it is provided by the wealthier nations of the world. The top contributors include Sweden, Norway, Denmark, the United States, Japan, and the Netherlands.

Local Leadership

Experts say quick and efficient disaster response requires locally trained people who are familiar with the language and culture of the country.

The Federal Emergency Management Agency (FEMA)

FEMA is working to reorganize itself by creating a clear chain of command and better plans to deal with the next large-scale American disaster.

Overview

"Our investigation revealed that Katrina was a national failure, an abdication of the most solemn obligation to provide for the common welfare. At every level—individual, corporate, philanthropic and governmental—we failed to meet the challenge that was Katrina. In this cautionary tale, all the little pigs built houses of straw."

—U.S. Congressional Report, "Katrina Response a 'Failure of Leadership.'"

"The only way to make disaster assistance work effectively is to have a plan before people act. Incident command training helps participants understand who they report to, what they will be asked to do and how to prepare to help."

—Judith Ward, coleader of the Mississippi disaster response team.

Deadly Events

Greg Henderson, a doctor from Washington, D.C., came to New Orleans in the late summer of 2005 for a medical convention. He barely escaped the city with his life. When Hurricane Katrina arrived on August 29, catching many residents off guard with its brutal power, Henderson hastily wrote an e-mail to his family back home: "The city has no clean water, no sewage system, no electricity, and no real communications." Henderson knew things would only get worse: "Bodies are still being recovered floating in the floods. . . . Infection and perhaps even cholera are anticipated major problems."[1]

For days, lawlessness reigned as armed looters roamed the watery streets of the city. As New Orleans sank into chaos and fear, millions of Americans watched on television as relief efforts, already slow to get under way, stalled, leaving thousands at the mercy of the monster storm. Henderson survived his ordeal, but over 1,800 people did not. Hurricane Katrina and its aftermath remain the costliest and one of the deadliest disasters ever to hit the United States.

"As New Orleans sank into chaos and fear, millions of Americans watched on television as relief efforts, already slow to get under way, stalled, leaving thousands at the mercy of the monster storm."

Three years later, on the other side of the world, armed gunmen attacked two hotels, a Jewish center, and a train station in Mumbai, India. At word of the unfolding killing and hostage-taking, Indian soldiers charged through the Taj Mahal hotel, trying to rescue as many people as they could. Despite their efforts, the terrorists murdered 195 people. "I think their intention was to kill as many people as possible and do as much physical damage as possible,"[2] said P.R.S. Oberoi, the chairman of one of the hotels that avoided attack.

According to reports, response to the attacks revealed weaknesses in the Indian security forces, including inexperienced soldiers, ineffective equipment, and poor communications. The terrorists, on the other hand, were well-coordinated and brutal: They entered the city by boat undetected and used cell phones and a global positioning system (GPS) to guide them on their mission.

Meeting the Challenges of Disaster Response

Hurricane Katrina and the Mumbai terrorist attacks are reminders that life is fragile and sometimes unpredictable. Hurricanes, tornadoes, wildfires, and tsunamis are but some of the natural disasters that can devastate an area, killing thousands and causing millions of dollars worth of damage. Other deadly crises can result from the spread of diseases such as cholera, HIV/AIDS, and malaria. An uncontained outbreak can grow into an epidemic, which may strain the resources of a nation and leave aid workers scrambling. People are also victimized by man-made catastrophes each year, including war, genocide, and terrorism, all of which kill scores of people and leaves countless more living in poverty and fear.

> **Although people cannot always control disastrous events, they can control, to some extent, how they respond to them.**

Although people cannot always control disastrous events, they can control, to some extent, how they respond to them. How they respond can affect the number of injuries and deaths that result from a catastrophe. Therefore, for aid agencies and most governments, primary goals include minimizing injuries and deaths through preparation, planning, and training. Much of this work must be done before disaster strikes so that when the worst happens, everything is in place and ready to go. In September 2008 aid workers in Louisiana, Florida, Alabama, Mississippi, and Arkansas were preparing for the aftermath of Hurricane Gustav. "Right now," said Jessica Vermilyea, the Louisiana disaster response coordinator for Lutheran Disaster Response, "we're just trying to focus on being prepared and making sure that people are prepared and getting out of town and securing what they do have.

". . . We'll deal with the aftermath of it, the recovery efforts, if need be. We hope that's not the case. But certainly it's always a possibility."[3]

Where Disasters Occur

Although disasters choose no favorites, patterns do emerge. For decades the continent of Africa has been ravaged by famine, disease, and war, and billions of dollars have been spent trying to stabilize many of Africa's 53 countries. The World Health Organization (WHO) reports that out of 100 disasters reported worldwide each year, about 20 occur annually in Africa. Yet 60 percent of all disaster-related deaths are suffered by Africans at least in part because of poor or nonexistent infrastructure, remote locations, and large, needy populations.

People of every continent are, at one time or another, challenged by disasters. According to a recent UN report, Asia was the continent most

This Navy flight crew is preparing to bring food to Banda Aceh, Indonesia, the hardest-hit area of the tsunami of 2004.

devastated by natural disasters in 2007 and 2008. The May 12, 2008, earthquake in China's Sichuan Province killed more than 70,000 people, 30,000 of whom were buried in the rubble of their homes or office buildings. Teams from the International Red Cross and Red Crescent Societies fanned out across the region saving as many lives as possible.

> **"The Office of the UN High Commissioner for Refugees houses and feeds those displaced by war and natural disasters. It assists approximately 33 million people in over 110 countries per year."**

The Americas have also experienced disaster: Deadly mudslides devastated Peru in 2004; flooding is a frequent problem in Canada. The United States, too, experiences its own share of calamity. From wildfires and earthquakes in California to flooding in the Midwest and South to severe seasonal changes in weather around the country, the United States struggles annually to keep its citizens safe and the damage to a minimum.

How Does the World Respond to Disasters?

Disaster response requires a combined effort by many different entities. Depending on the location and severity of the disaster, the response may involve local police and fire departments, hospitals, private relief organizations, government agencies, and sometimes even branches of the military. Such efforts take a great deal of planning and coordination. Identifying priorities; assessing needs; getting food, water, and medical supplies to victims; and having trained personnel to do all of this work is just part of what disaster response entails.

Disaster response also takes a great deal of money. In 2008 alone, the UN Central Emergency Response Fund (CERF) spent more than $140 million on food aid, more than $57 million on health, and nearly $38 million on water and sanitation in response to natural disasters and armed conflicts around the world.

Nongovernmental Organizations

Private organizations that deal with disaster relief and response are known as nongovernmental organizations, or NGOs. The International Federation of the Red Cross and Red Crescent Societies is probably the most well known of these groups. It provides feeding stations, shelter, cleaning supplies, comfort kits, first aid, blood and blood products, food, clothing, emergency transportation, rent, home repairs, household items, and medical supplies in response to disasters occurring around the world. The Red Cross is just one of hundreds of NGOs. Others include Médecins sans Frontières (known in the United States as Doctors Without Borders), which provides emergency medical care; Children's Disaster Services (CDS), which provides child care in shelters and disaster assistance centers so that parents can continue working; and the Friends Disaster Service (FDS), which helps with cleanup and rebuilding assistance to survivors of disasters.

Then–UN secretary general Kofi Annan spoke in 2000 about the importance of NGOs, many of which are supported by private donations. Calling them "our best defence against complacency, our bravest campaigners for honesty and our boldest crusaders for change,"[4] Annan reminded the world of what millions of disaster victims witness every day: NGOs can mean the difference between life and death.

The Humanitarian Relief Supply Chain

Supplies and services reach disaster victims through what is known as the humanitarian relief supply chain. "The humanitarian relief supply chain aspires to deliver the right supplies to the right people, at the right place, at the right time, and in the right quantities,"[5] one expert writes.

Thousands of tons of food and water get shipped each year to devastated locations, as do temporary shelters like tents or trailers. In 2008, for instance, the UN Central Emergency Response Fund spent almost $219 million on food, shelter, water, and sanitation for victims of natural disasters and armed conflicts. When Yvan Sturm arrived in the dry central African nation of Chad in 2003 to help Sudanese refugees fleeing war in their country, getting water to the people was his first concern. Sturm, who works for the United Nations High Commissioner for Refugees (UNHCR), says, "What preoccupied me most was water. It's like gold in this region. The wind and sand will dry you up within two hours."[6]

This type of help is by no means limited to developing nations such as Chad. Severe flooding in Bonita Springs, Florida, in 2008 forced more than 500 people to evacuate their homes and take shelter in the gymnasium of a local community center. There, Red Cross volunteers provided evacuees with cots, clothing, and food until they could return to their homes.

Determining what supplies are needed, when, and where is a complicated task. Complications are magnified in large-scalc disasters when communication and transportation may be poor to begin with and where multiple agencies and governments are trying to provide assistance.

Many Disasters, Limited Resources

The frequency of disasters, natural and man-made, has severely strained international and local disaster operations in recent years. The Office of the UNHCR houses and feeds those displaced by war and natural disasters. It assists approximately 33 million people in over 110 countries per year. The cost of such operations total at least $1.2 billion a year, much of it coming from UN member nations. While that sounds like a huge amount of money, it is never enough.

Zimbabwe, in southern Africa, suffered an epidemic of cholera beginning in late 2008. The bacterial disease, which causes severe diarrhea and can kill, is spread through contaminated drinking water. By late 2008 the UN reported 16,000 cases of cholera; deaths were in the thousands and rapidly climbing. More than 500 national and international Médecins sans Frontières (MSF) staff members worked to identify new cases and to treat patients in need of care, but they were quickly overwhelmed. "The scale and the sheer numbers of infection especially in [the city of] Harare is unprecedented,"[7] said one doctor.

How Does the United States Respond to Natural Disasters?

Each year, federal, state, and local organizations work together to reach the victims of floods, hurricanes, heat waves, earthquakes, and wildfires. In many cases, this national support system is carried out seamlessly, as victims are given a safe place to sleep, food to eat, and sometimes a loan to get them back on their feet. But disaster relief is costly. American tax-

This rescue helicopter lifts a person who was trapped on the roof of a house due to the flooding that destroyed New Orleans during Hurricane Katrina.

payers spent between $3 billion and $6 billion in 2007 saving lives and rebuilding communities.

Despite the enormous amount of money spent responding to calamities, the process can be sometimes less than perfect and at other times, say critics, borders on negligent. One million people (the majority of whom lived in New Orleans) were displaced by flooding in the aftermath of Hurricane Katrina. Thousands lost their homes and had limited access to food, water, and shelter. In this instance, experts say, local, state, and federal governments failed to respond adequately to the needs of those affected by the disaster.

Working as a volunteer with AmeriCorps in Biloxi, Mississippi, in 2007, 24-year-old Tom Halloran discovered the joys and the limitations of helping the victims of crises. On one hand, he says, the work "made me appreciate the gift of giving." At the same time, Halloran and his coworkers sometimes became frustrated by the process: "There are so many obstacles in the way, mainly red-tape and organizational obstacles. There would be days where we would have to stop midday because we were not going to get the funds approved to purchase hardwood flooring or paint."[8]

Is the United States Prepared for Man-Made Disasters?

September 11, 2001, marks the worst terrorist attack in U.S. history. But most experts agree that the next major assault will not be done with airplanes. Instead, the fear of an attack using nuclear weapons has forced the United States to keep a closer eye on who has such deadly weapons and where they are located. Globally about 30,000 nuclear weapons are held by various countries: India, Pakistan, Israel, China, Russia, and the United States, among others. Others, most notably Iran and North Korea, may be are on the verge of acquiring or making them.

A Senate panel held in June 2008 discussed the measures being taken to prepare for a nuclear blast in an American city. While all in attendance agreed that a nuclear attack would be catastrophic and likely kill thousands and injure thousands more, officials said that measures were being taken to minimize the potential loss of life. Assistant Health and Human Services secretary for preparedness and response Craig Vanderwagen testified that hospital infrastructure had improved enormously since the 2001 terrorist attacks. He stated that 87 percent of hospitals in the United States are prepared for mass casualties.

> **"Experts warn that biological or chemical agents such as smallpox, ebola, anthrax, mustard gas, and ricin would also pose grave danger to the general population if these came into terrorist hands."**

As for the radiation that would result from a nuclear attack, Vanderwagen was equally confident that emergency workers would be ready: "We have the ability to decontaminate over 400,000 people within three hours on a nationwide basis,"[9] he said. Such preparation, along with an enhanced security budget and 20,000 trained military personnel, provides some hope that the country is ready to respond in the face of such a potential catastrophe.

Biological and Chemical Threats

Experts warn that biological or chemical agents such as smallpox, ebola, anthrax, mustard gas, and ricin would also pose grave danger to the general population if these came into terrorist hands. These deadly substances, often invisible to the eye, could kill millions throughout the country. Americans learned about the potential for such an attack in 2001, when anthrax powder sent through the mail killed 5 people and sickened more than 50 others.

Federal agencies, including the Centers for Disease Control and Prevention (CDC), are taking no chances. Over the last decade, the CDC and the U.S. Department of Health and Human Services have worked to educate the American public about such dangers, while also attempting to develop new vaccines to protect citizens and to ramp up response efforts.

For Peggy Hamburg, a former New York health commissioner, a quick response in the aftermath of chemical or biological attack is everything. "The key to preserving lives and reducing the overall human and economic cost of such an attack will be rapid recognition of what's going on and mobilisation of response, and that's why putting in place a public health system that really enables rapid detection is key to our ability."[10]

The Politics of Disaster Response

One major obstacle to providing timely disaster response is politics. When leaders act swiftly and competently to aid their nation's disaster victims, they are often hailed by their own people and around the world. Sometimes, though, leaders insist that their countries remain closed to outside influence. These countries may accept aid only reluctantly, even when overwhelmed by famine, war, or the destructive forces of nature, "Unfortunately," says former Federal Emergency Management Agency official Barry Scanlon, "it seems that political situations in some areas of the world help people make the wrong decisions, and keep them from receiving the aid and helping their people when they obviously need it."[11]

In 2008 President Robert Mugabe of Zimbabwe, in an effort to retain his hold on the country, refused international aid during his country's cholera crisis. For Mugabe, any help from the outside world was a sign of weakness as well as an opportunity for relief groups to witness the brutality of his government firsthand. His refusal endangered more lives. "Infections are still climbing," said WHO spokesman Gregory Hartl in December 2008, "and with the rainy season on the way the situation could get worse."[12]

> **Improvements in disaster response are often slow in coming, and not all experts agree on how best to meet the challenges posed by natural and man-made disasters.**

Burma's Cyclone Nargis killed at least 130,000 people in May 2008. Burma, renamed Myanmar by its oppressive government, had also shunned all help from the outside world. Yet the overwhelming nature of the storm convinced them to allow humanitarian groups in to help rebuild schools, roads, and homes. Thus far $240 million has been donated to help victims of the cyclone.

How Can Disaster Response Be Improved?

Improvements in disaster response are often slow in coming, and not all experts agree on how best to meet the challenges posed by natural and man-made disasters. Speaking before a House of Representatives committee on February 25, 2009, Department of Homeland Security (DHS) secretary Janet Napolitano set her priorities: building more effective partnerships between federal, state, and local governments; using science and technology to better prevent and respond to disasters; and unifying a department whose reputation was tarnished during Hurricane Katrina.

The UNHCR has also been evaluating its disaster response capabilities. In a 2005 report it suggests that the main part of its mission is in danger of being lost. The agency, reads the report, "has a tendency to behave as though its primary purpose is . . . to create reports, arrange staff movements, and keep itself funded," instead of "protecting and assisting refugees."[13] Thus, one goal is to place the emphasis back on the victims of catastrophic incidents. One way of doing this may be utilizing technology to better communicate with those in need and get them the supplies they require more quickly.

Beyond that, says national security expert Stephen Flynn, organizations must provide the basics. "Acts of terror and disasters cannot always be prevented," he writes, "but they do not have to be catastrophic. The key is being willing to invest in things that are not particularly sexy, such as public health, emergency planning and community preparedness."[14]

How Does the World Respond to Disasters?

“It was truly amazing to see how devoted the hospital staff in the affected area reacted. I have never seen so many people act so selflessly to ensure the safety of so many.”

—Charles Skowronski, an American who lives in China and survived a massive earthquake there in May 2008.

“I’m angry that we’ve had to stay here for so long. . . . I’m angry at the government and the NGOs [nongovernmental organizations] because they just come here and promise to build us a house, and then we never see them again.”

—Anwar, a resident of Banda Aceh, Indonesia, who survived a massive tsunami in 2004.

Getting the Work Done

David Croft works on the front lines of some of the world’s many disasters. He is a logistician for Médecins sans Frontières (MSF), an international organization that sends volunteer doctors and nurses to disaster areas. Croft drives trucks, fixes generators, and hauls bags of food—doing whatever needs to be done to keep relief efforts going.

Logisticians, like the doctors and nurses they serve, are an essential part of any disaster response effort; they make the seemingly impossible happen. When supplies are needed, they get them; when equipment is broken, they fix it; when local aid workers fall asleep, they wake them up and insist the job get done. "In Sierra Leone, I needed a clinic in ten days," recalls one nurse, "and they built the clinic in ten days from scratch—fence, walls, buildings, wells, latrines."[15]

Global disaster response requires millions of dollars and the international coordination of massive relief supplies, including food, medicine, and shelter for the victims of the tragedies. It also depends on thousands of individuals like Croft, skilled professionals willing to risk their own safety by rushing to the scene of an emergency when a tsunami crashes onto a beach, or an airplane crash-lands in the Hudson River, or a country goes to war.

Coordination and Communication

Disaster operations require coordination and cooperation. Various relief organizations and governments, which can number in the hundreds or thousands during a large-scale disaster, must work together to ensure that a population's needs are met—quickly and efficiently. Remote locations, bad weather, or both can make delivering large quantities of supplies and hundreds of workers challenging at best. Good intentions can also go awry.

People around the world responded generously to calls for donations after the December 2004 Indian Ocean tsunami that devastated parts of Indonesia, Sri Lanka, Thailand, and India, killing 225,000 people and leaving millions homeless. But poor communication and coordination resulted in mounds of unneeded and unusable supplies. A 2005 analysis of the tsunami response noted,

> In Sri Lanka, unwanted aid accumulated at government buildings, aid agencies, and refugee camps. Many weeks after initial appeals for water, significant numbers of boxes of bottled water continued to arrive after water and sanitation services were restored. . . . Winter jackets, winter tents, expired cans of salmon, cologne, high-heeled stiletto shoes, and sequin-studded black evening dresses were sent by well meaning people and organizations.[16]

Communication is therefore a vital tool in saving lives and getting the right supplies where they are needed most.

No single system exists to connect the nations and the many organizations trying to help the victims of catastrophe. Thus, precious time is often spent simply trying to communicate the needs of refugees, disaster victims, and the relief workers themselves. Such challenges often lead to the NGOs going it alone with little help from other groups. "Despite coordination attempts," says emergency management expert Damon P. Coppola, "these responding agencies tend to work independently and in an uncoordinated manner."[17]

Quick Response Can Save Lives

The first hours and days after disaster strikes are crucial in minimizing the loss of human life. Searching for survivors and helping the wounded are among the first tasks on any disaster scene. But within a short time, workers on the scene begin planning for food, shelter, and other basic needs of those affected by the disaster.

After the 2004 tsunami hit, formal relief operations got under way in Indonesia with the arrival of Colonel Tan Chuan Jin of the Singapore Armed Forces (SAF). Tan's job was to organize the response. His first priority was determining the needs of the Indonesian people and breaking the rescue operations into smaller, manageable pieces. Banda Aceh, one of the hardest hit areas, quickly became a focal point.

Tan's job was made more difficult by the fact that the tsunami had killed 35 percent of Indonesian Red Cross workers. Despite this hardship, within a matter of hours Tan had dispatched three landing ships and a fleet of Chinook helicopters to the area. The massive operation bringing food, workers, and temporary housing also utilized the Singapore Red Cross and the YMCA of Singapore.

> **"Logisticians, like the doctors and nurses they serve, are an essential part of any disaster response effort; they make the seemingly impossible happen."**

Arriving doctors feared that cholera, a disease spread by contaminated water, might break out in the immediate aftermath of the tsunami. Therefore, the chlorine tablets they brought with them were quickly used to disinfect the water; victims were also given special salts to fight diarrhea. Another disease-fighting precaution was the fast burial of the dead in mass graves, although with tens of thousands of bodies this process sometimes took days or weeks.

How Geography and Climate Affect Relief Efforts

Responding to disasters in remote parts of the world or in places that experience extreme weather can be especially challenging. Relief efforts during the 2004 tsunami were complicated by the geography of Indonesia, which consists of thousands of islands. Because of this, UN undersecretary Jan Egeland told the Indonesian government that the first response would not come from the outside world: "It is your own people through local communities, authorities, and civil society that are doing all the lifesaving for the time being."[18]

> **"The first hours and days after disaster strikes are crucial in minimizing the loss of human life."**

This was also the case during an October 2005 earthquake in Pakistan, which killed more than 80,000 people. Early response to the quake was slow because of the mountainous terrain. Three weeks after the event, at least 2 million people were still in need of immediate help.

By January 2006 winter storms and torrential rains made response efforts even more precarious, but, in the biggest helicopter operation in its history, the World Food Programme delivered 9,370 tons (8,500 metric tons) of food and 76 mobile warehouses to affected areas. Despite challenges of terrain and weather, thousands of lives were saved.

Where the Money Comes From

Aiding the victims of calamity would be virtually impossible without millions of dollars of aid money that pours in from around the world each year. In 2008 relief organizations distributed nearly 17,000 tons (15,422 metric tons) of food to 891 million victims of disasters throughout the world. In 2009 supplies, medical care, and personnel will cost the International Committee of the Red Cross 12 percent more than in 2008, at least $464 million. Although it is one of the largest, the Red Cross is only one of at least 40,000 NGOs vying for funding around the world.

While some of the funding for disaster response and relief comes from private donations, most of it is provided by the wealthier nations of the world. In 2008 the top three donors of humanitarian aid were Scandinavian countries: Sweden, Norway, and Denmark. Close behind were the United States, Japan, and the Netherlands. Although exact figures were not available, these and other countries spent billions of dollars helping the poorest and neediest nations in the world.

> **Although careful not to take sides in conflicts, aid workers must always be on their guard for danger.**

With enormous amounts of money and resources being spent each year on disaster relief, concerns sometimes arise about how funds are spent or where they actually go. In 2007 a U.S. agency overseeing the rebuilding of postwar Iraq reported on mismanagement and corruption in the country. Millions of dollars in aid were being stolen, in some cases by Iraqi officials or contractors. Those who needed the relief most were simply not getting it. "Millions of Iraqis have been forced to flee the violence," said Oxfam director Jeremy Hobbs. "Many of those are living in dire poverty."[19]

Concerns About the Safety of Relief Workers

Ensuring the safety of relief workers is a high priority for organizations that respond to disasters. In many cases, relief efforts take place in or near war zones, and this sometimes delays the response. Relief efforts have been slowed by bloody conflicts or by reluctance to put employees in harm's way in countries such as Somalia, Sudan, Sri Lanka, and Afghanistan. At times, aid workers have even become targets themselves. Although careful not to take sides in conflicts, aid workers must always be on their guard for danger.

In recent years relief workers have frequently become the targets of violence. In 2008, 36 were murdered in the African nation of Somalia, 11 were killed in Darfur (Sudan), and 4 died in Chad. In the early months of 2009, 2 aid workers were killed in Pakistan. Between 1997 and 2005, violence against NGO employees worldwide increased 92 percent.

Thirty-three relief workers died in Afghanistan in 2008, including

34-year-old Gayle Williams, who worked for SERVE Afghanistan, a British aid organization. Williams was shot by a gunman on a motorcycle while she walked to work. "We employ low-profile security and prefer not to use armed guards," said a fellow aid worker. "We are a much softer target than the military. Unfortunately I don't think this will be the last casualty."[20]

During warfare between Israelis and Palestinians in January 2009, the UN halted all relief supplies it was sending to Gaza after one of its truck drivers was killed by tank fire. "We've been co-ordinating with them (Israeli forces) and yet our staff continue to be hit and killed,"[21] said a UN spokesman.

Housing Refugees

Typically a primitive camp is set up to house thousands of desperate and needy people. Hydrologists (water experts) and engineers are on hand to determine whether water is available nearby and to lay water pipes and dig latrines. Doctors tend to the sick and the dying. In just a few short months, a small grouping of tents can grow into a small city with a population of tens of thousands. Refugee camps are meant to provide temporary housing until victims can safely go home. But temporary shelters can become more permanent dwellings, especially as countries remain at war and people have nowhere else to go.

Life in refugee camps provides little privacy. Families may be given a tent, which can house four to six people, and a small plot of land on which they may grow crops to feed themselves or sell to others. But the camps, no matter how well run or organized, cannot replace the home of a displaced person.

> **"Relief workers must treat not only the bodies but the minds of those whose lives have been turned upside down by disaster."**

Forty-five-year-old Adelia Simbe remains haunted by the flooding in 2006 that drove her from her home in Zambezia, one of Mozambique's poorest provinces. The rising waters swallowed all she owned and destroyed her ability to farm rice, beans, and corn.

Today she lives in a refugee camp in a hut made of sticks and grass. Having a place to sleep does not always make life easier. "It is tough in the camps," she says. "There is not much food and we don't have pots and plates. We struggle with getting water and now it is hard to go to the river because of the crocodiles."[22]

Healing Scars

One of the most challenging jobs in disaster response is tending to the psychological wounds resulting from a natural or man-made disaster. Extreme stress caused by fear or the loss of loved ones takes a toll on refugees and accident victims, many of whom need counseling to help them work through their overwhelming and painful experiences.

Children are particularly vulnerable to psychological damage during a disaster. Psychologists and other health care responders often persuade young patients to talk about their experiences. When they do not want to talk, other kinds of therapy, including making music and drawing, are sometimes used to allow the children to share their experiences.

For adults, dealing with the aftermath of a horrible experience can be even harder. Since 2003 the Darfur region of western Sudan has experienced the genocide of hundreds of thousands of people. Murder and mutilation have become commonplace. As in most wars, women become easy targets, and aid workers spend much of their time treating girls and women who have been brutally raped.

"The first priority after a rape is medical care," says Charlotte Tabaro, a relief worker. "A woman who has been raped must get to a health centre as quickly as possible. But medicines can only treat the body. The victims of these attacks bear invisible, psychological wounds."[23] Such wounds can last a lifetime, and relief workers must treat not only the bodies but the minds of those whose lives have been turned upside down by disaster.

How Does the World Respond to Disasters?

“This United Nations is where the world turns to confront some of its greatest challenges . . . where the hopes of the world for a better future ultimately rest.”

—Gordon Brown, “On the World Economy,” United Nations, September 26, 2008. www.un.org.

Brown is prime minister of Great Britain.

“Even when the UN has been used solely as a mechanism to deliver resources to NGOs, the system has been slow, unwieldy, inefficient and unresponsive to the needs of affected communities.”

—Parliament, House of Commons International Development Committee, (Malcolm Bruce, Chair), Great Britain, “Humanitarian Response to Natural Disasters,” volume II, London: TSO Shop, 2005-06.

Parliament is the chief legislative body for Great Britain and its overseas territories.

* Editor’s Note: While the definition of a primary source can be narrowly or broadly defined, for the purposes of Compact Research, a primary source consists of: 1) results of original research presented by an organization or researcher; 2) eyewitness accounts of events, personal experience, or work experience; 3) first-person editorials offering pundits’ opinions; 4) government officials presenting political plans and/or policies; 5) representatives of organizations presenting testimony or policy.

"While the Japanese government's slow and confused response to the [Kobe earthquake] disaster was undeniable, around 1.5 million individuals responded rapidly by taking on a considerable proportion of relief activity."

—Alpaslan Özerdem and Tim Jacoby, *Disaster Management and Civil Society*. London: I.B. Tauris, 2005.

Özerdem teaches at the University of York. Jacoby is a member of the Institute for Development Policy and Management at the University of Manchester.

"From the perspective of the overall [hurricane] disaster relief system whether the state was too slow asking or the federal government too slow in giving is ultimately not as important as is the fact that the net results were unsatisfactory."

—James F. Miskel, *Disaster Response and Homeland Security*. Palo Alto, CA: Stanford University Press, 2008.

Miskel is a writer and expert on disaster relief.

"Just tell the people the truth. If you're going to build the houses in three months time say so but do not promise them next week, for example. The promise is a debt, so if you promise something to them they consider that as a debt you have to pay. That's why they're angry."

—Eddy Purwanto, interview, "Slow Rebuilding Process Frustrates Tsunami Victims," *All Things Considered*, National Public Radio, December 22, 2005. www.npr.org.

Purwanto is deputy director of Aceh's Reconstruction Agency.

"A systematic process for responding to international disasters has begun to emerge. . . . What was only 20 years ago a chaotic, ad hoc reaction to international disasters has grown with astounding speed into a highly effective machine."

—Damon P. Coppola, *Introduction to International Disaster Management*. St. Louis, MO: Butterworth-Heinemann, 2006.

Coppola is a systems engineer and a senior associate with Bullock and Haddow LLC, a disaster management consulting firm.

"We offered a lot of moral support to Palestinian doctors [during war in Gaza], as well as physical support. We gave them a chance to rest a little and take a breather from rigors of the job at hand. When I arrived, the director of the ICU was working non-stop."

—Said Abu Hasna, interview, "Gaza: From Qatar with a Mission," International Committee of the Red Cross, January 27, 2009. www.icrc.org.

Abu Hasna is a Red Crescent doctor from Qatar.

"The health center, by a miracle of nature, has survived [the Peruvian earthquake]. But inside, the staff are like zombies. Over a week, while they struggled to get back on their feet, the number of consultations tripled. The doctor has no mental strength left."

—LuisE, "Diary from Peru: Helping Another Kind of Displaced People," Médecins sans Frontières Tour Blog, August 26–28, 2007. http://msf.ca.

LuisE worked with Médecins sans Frontières in South America.

"Steps must be taken to ensure that the violently wounded can access lifesaving medical care. This access must not be conditioned on any considerations, political or otherwise, except need."

—Christophe Fournier, "The Humanitarian Situation in Haiti," Médecins sans Frontières to the UN Security Council, "Arria Formula" meeting, April 8, 2005. www.doctorswithoutborders.org.

Fournier is the current president of Médecins sans Frontières/Doctors Without Borders.

"Coping with and reducing disasters so as to enable and strengthen nations' sustainable development is one of the most critical challenges facing the international community."

—World Conference on Disaster Reduction, "Hyogo Declaration (excerpt)," January 18–22, 2005. www.unisdr.org.

The conference, held in Kobe, Japan, was partly sponsored by the United Nations.

How Does the World Respond to Disasters?

- The UN Central Emergency Response Fund, established to enable more timely and reliable humanitarian assistance to people affected by natural disasters and armed conflicts, gave **$428.8 million** to 55 countries in 2008.

- The United States and Switzerland are the largest contributors to the International Committee of the Red Cross (ICRC); contributions from those nations along with the European Union, Australia, Canada, Japan, and New Zealand make up about **80–85 percent** of the ICRC's annual budget.

- Food aid, health activities, and water and sanitation programs make up **58 percent** ($464 million) of the 2009 budget of the ICRC, a **12 percent** increase over 2008.

- The ICRC employs **2,000** professional employees, with roughly 800 workers in its Geneva headquarters and 1,200 people in the field.

- Médecins sans Frontières (MSF) provides medical aid in **60 countries** around the world, often in conditions of war.

Reported Disasters on the Rise Worldwide

There has been a dramatic rise in natural disasters over a 33-year span. While the increase may reflect improvements in reporting such catastrophes, the spike in occurrences has led many experts to suggest improvements in disaster response around the world.

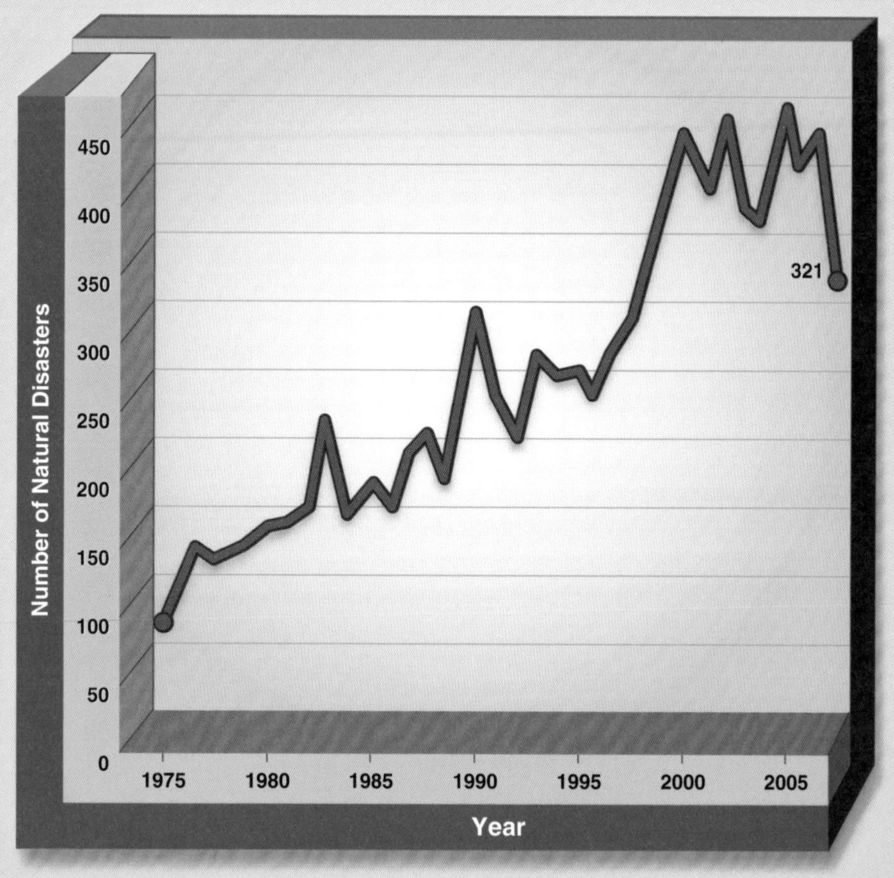

Source: United Nations, International Strategy for Disaster Reduction, "Time Trend of Reported Natural Disasters, 1975–2008." www.unisdr.org.

- According to the Federal Emergency Management Agency (FEMA), the first **72 hours** after a disaster are critical, and each person should be prepared to be self-sufficient—able to live without running water, electricity and/or gas, and telephones—for at least three days following a disaster.

The 2004 Tsunami

The largest tsunami disaster ever recorded hit India along with other South Asian countries on December 26, 2004. The illustration below shows the areas aided by the Lutheran World Service India (LWSI), one of dozens of organizations charged with helping devastated areas around the world. Among other things, the LWSI provided fishermen's livelihood kits, temporary shelters, and counseling.

Source: Lutheran World Service India, "Projects," 2007. www.lwsi.org.

- Eight schools in the United States currently offer **emergency management–related doctorate programs.**

The Humanitarian Disaster in Darfur

Not all disasters are caused by nature. In the case of the Darfur region of Sudan in northeastern Africa, a humanitarian disaster was created when militias allied to Sudan's government began committing acts of violence against those citizens living in the western region of the country. This chart, based on eyewitness reports, illustrates the percentage of Darfur's people who have witnessed or experienced various acts of violence. Such statistics provide relief workers with a better understanding of the scope of the crisis as they work to provide food, shelter, and counseling to the victims.

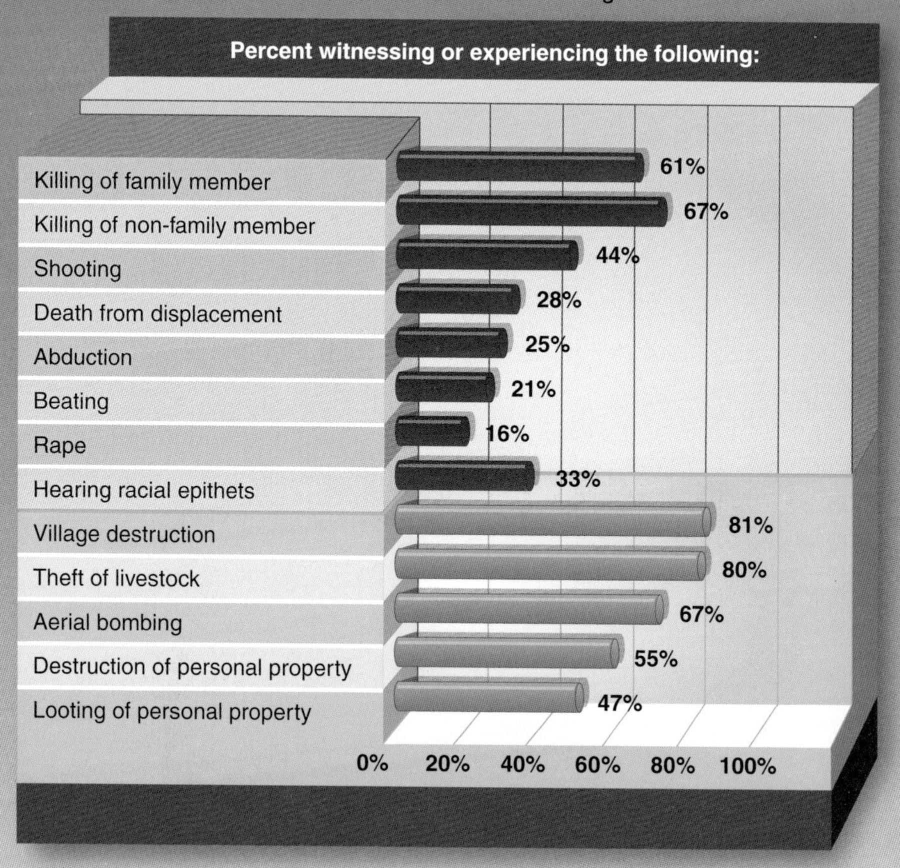

Source: U.S. Department of State, "Documenting Atrocities in Darfur," 2007. www.state.gov.

- The UN Children's Fund (UNICEF) provides postdisaster shelter, water, and medical supplies in over 150 countries and territories; its total income for 2006 was **$2.78 billion**.

The 2008 Earthquake in China

On May 12, 2008, a magnitude 8 earthquake shook southwest China. The map outlines the initial reports that came in on May 15, just 3 days after the earthquake. Later reports have much higher numbers. According to UN numbers in July 2008, 46 million people (this includes dead, missing, and displaced) were affected by the disaster. As of August 2008 the United States had given more than $48 million in humanitarian funding to the Chinese.

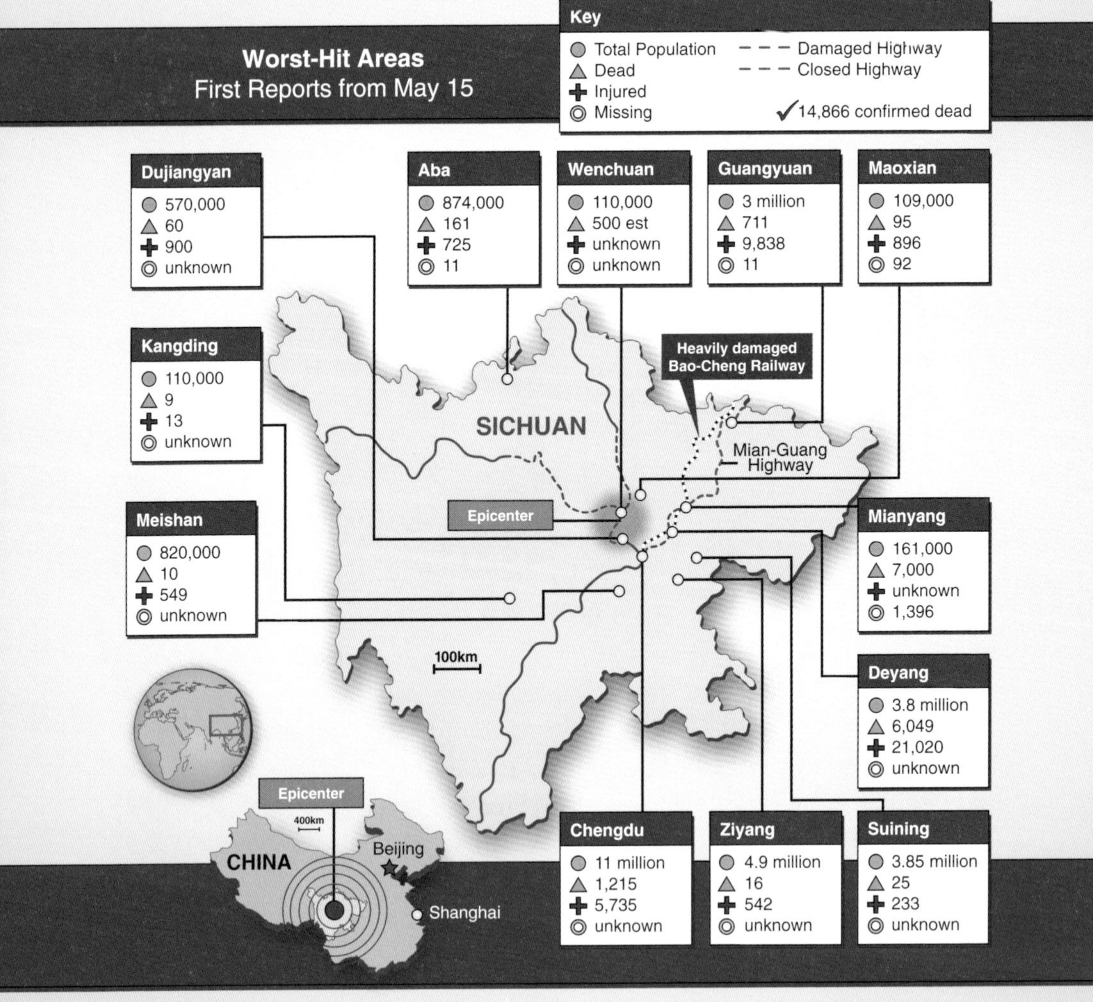

Sources: China Today, "Strong Earthquake Relief to China," May 15, 2008. www.chinatoday.com; USAID, "USAID Provides Earthquake Relief to China," August 8, 2008. www.usaid.gov.

- World Food Programme, a food aid branch of the UN, provides food to **90 million people** per year, many of them survivors of disasters.

How Does the United States Respond to Natural Disasters?

“Even given the failure to plan and prepare for a storm like Katrina, the story might have been different had the federal, state, and local response to the storm not been so inept.”

—Thomas A. Birkland, professor of public administration and policy at the University at Albany, SUNY.

“[The federal government] really hit the ground running. They’re working very hard to get all the equipment and supplies here that we need.”

—Steve Beshear, governor of Kentucky.

A winter storm in January 2009 left 1.3 million Americans in the dark from the Ozark Mountains to Appalachia. The state of Kentucky was especially hard hit: Frozen power lines and downed trees caused havoc and killed dozens of people. Kentucky governor Steve Beshear called the ice storm the biggest natural disaster the state had ever suffered.

Almost immediately after the storm hit, Beshear ordered 4,600 members of the Kentucky National Guard to go door-to-door, providing food and water or evacuating freezing victims to local shelters. The shelters, located in gymnasiums, schools, and recreation centers, soon filled with cold and hungry people. Beshear spoke bluntly about his state’s priorities: "With the length of this disaster and what we're expecting to be a multi-day process here, we're concerned about the lives and safety of our people in their homes."[24]

The Federal Emergency Management Agency (FEMA) coordinated rescue efforts, and organizations like the American Red Cross served hot meals to thousands. Less than a week into the crisis, spirits began rising with the temperatures, and relief efforts were deemed a success. As this example illustrates, disaster response in the United States can go smoothly, but it does not always work that way.

The Federal Response to Hurricane Katrina

Millions of Gulf Coast residents had a far different experience on August 29, 2005, when Hurricane Katrina swept into the Gulf of Mexico, barreling toward Louisiana and Mississippi. Before long, 18- to 25-foot (5.5 to 7.5m) storm surges were topping New Orleans's levees and flooding the area. In a matter of hours, 80 percent of the city was underwater. Thousands of citizens were stranded, desperate for government assistance.

Two days after the storm, the president announced that FEMA had moved 25 search-and-rescue teams into the New Orleans area. Yet millions of Americans watched in horror as 24-hour news coverage brought the desperation, pain, and ugliness of New Orleans into their homes: Thousands of starving and thirsty people sat stranded at the New Orleans convention center, wilting in the late summer sun; toxic chemicals poisoned the water; and gangs of criminals began roaming the streets. Three days after the storm struck, as buses finally arrived to cart victims to safety, hundreds of people were already dead. Eventually, at least 1,836 people along the gulf died as a result of the hurricane and in the floods that followed, making it one of the deadliest storms in U.S. history.

"Before long, 18- to 25-foot storm surges were topping New Orleans's levees and flooding the area. In a matter of hours, 80 percent of the city was underwater."

A 2006 congressional report totaling 600 pages slammed the federal government's response to Hurricane Katrina, putting much of the blame on Homeland Security secretary Michael Chertoff and FEMA director Mike Brown: "If 9/11 was a failure of imagination then Katrina was a failure of initiative," read the report. "It was a failure of leadership. In this instance, a blinding lack of situational awareness and disjointed decision making needlessly compounded and prolonged Katrina's horror."[25]

FEMA

FEMA, created in 1979 by then-president Jimmy Carter, is charged with protecting the lives and property of Americans by providing food, medicine, and temporary housing to those affected by disaster. With a 2008 budget of $5.8 billion, FEMA works as a failsafe when state or local resources and personnel are overwhelmed by the scope and size of a disaster.

The agency, though, can only swing into action once a state governor declares a state of emergency and formally requests the support of the federal government. Once that occurs, FEMA's 2,600 disaster assistance workers and 4,000 standbys coordinate response efforts with state and local officials either from its headquarters in Washington, D.C., or one of its 10 regional offices around the country. FEMA workers will then arrive at the scene of a disaster to track and distribute supplies, communicate with contractors and NGOs to provide faster relief, and work with insurance companies to begin rebuilding efforts.

"FEMA workers will . . . arrive at the scene of a disaster to track and distribute supplies, communicate with contractors and NGOs to provide faster relief, and work with insurance companies to begin rebuilding efforts."

A recent brochure touted the focus and flexibility of the agency, but over the years FEMA's report card has been decidedly mixed. Hurricane Andrew in 1992 left 250,000 people homeless in Florida and Louisiana. FEMA was criticized for its slow response and inability to feed and house those in need. The agency again angered Americans in 2006 when it took 3 days to respond to a heavy snowstorm in Buffalo, New York.

FEMA defended itself by saying that the state's governor had not asked for help.

In 2006, less than a year after FEMA received harsh criticism for its handling of Hurricanes Katrina and Rita, the U.S. Congress passed the Post-Katrina Emergency Management Reform Act, which the agency says created a "New FEMA." While acknowledging past mistakes, FEMA claims to have implemented operational changes that make it better able to meet the needs of families and communities in times of crisis. These include establishing a clear chain of command and developing a response plan before moving into action.

State Relief

In the hours immediately following an earthquake, blizzard, or mudslide, emergency management responsibility falls to the state in which the calamity occurred. Each state has an agency dedicated to disaster response and a budget—usually in the millions of dollars—to pay for relief efforts.

In Florida, the Division of Emergency Management routinely performs exercises to test the state's ability to respond to disasters in a flexible and timely manner. The division also serves as a liaison between federal and local response agencies and advises Florida's governor whether to declare a state of emergency and apply for federal funds.

As with other similar agencies across the United States, Florida's focus is on taking care of disaster survivors and providing responders with the support they need to perform in crisis situations. This means that the state is in charge of purchasing the latest equipment for police, firefighters, medics, and emergency medical technicians (EMTs).

Federal or State Responsibility?

Hurricane Katrina put a spotlight on the often complicated relationship between states and the federal government during moments of crisis. In a press conference in 2005, Chertoff made the federal government's view clear: "Our constitutional system really places the primary authority in each state with the governor."[26] And while this may be true in theory, the reality is less clear-cut, as state and local officials often argue over where the ultimate responsibility for disaster relief lies.

In early 2009, 6 months after Hurricane Ike rolled through Texas, a dispute arose that illustrates the sometimes convoluted disaster response system. Some 350 businesses that had provided portable toilets and showers, transportation, and other assistance to victims of the disaster were angry and complaining. They had not been paid for their services, they said, and estimated the total amount owed them to be over $134 million.

Texas governor Rick Perry blamed FEMA for not paying up; FEMA insisted that Texas first pay the businesses and then apply for reimbursement from the federal government. One business owner said that he could not get a straight answer about when his company would be paid; others in the same situation said they would be reluctant to work with the state again. The state government feared that the dispute between the federal and state agencies would damage relief efforts going forward. "The state of Texas depends on these vendors to provide much-needed services in times of emergency," said a spokeswoman for the governor, "and we want to be able to do business with them in the future."[27]

> **"Apart from the trained professionals and elected officials, local disaster response is often a grassroots operation. Town officials may work with church groups or accept donations of food, clothing, and money."**

Fighting Wildfires

The state of California spent more than $305 million fighting wildfires in 2008, $236 million more than its lawmakers had set aside. Hiring enough firefighters and then paying for the enormous amount of equipment needed to douse the flames is expensive. State lawmakers are forced to find new ways to pay for their disaster response bills, often by cutting services or health-care benefits for citizens, as was proposed in 2008 by California governor Arnold Schwarzenegger.

Although the financial responsibility for fighting wildfires is divided three ways—between federal, state, and local governments—strain on the state is great. Disasters like wildfires can mean potential loss of life and damage to a state's infrastructure—roads, bridges, and highways. At the state level, disaster response can overwhelm even a large budget.

One proposed solution is to keep a community safe by letting the wildfires burn in unpopulated areas. The strategy may save the state money while also helping restore the forest's natural cycle. Still, according to author Benjamin Wisner, "Disasters occur when hazards meet vulnerability."[28] In California, as in other parts of the United States, battling that vulnerability typically begins at the local level.

Local Relief

Most small towns and large cities across the country have disaster plans in place for saving lives and keeping their communities intact. These plans may include where to shelter victims, how much food might be needed, and what kinds of medical care are available.

A spring tornado in Macon, Georgia, in 2008 put those plans to the test. It began with local police patrols checking in on people, making sure they were okay. Soon after, city crews were out helping clear trees and the 450 tons (408 metric tons) of debris left by the storm. In just a few days, Macon's budget was taking a hit. With city workers on duty nearly around the clock, overtime costs alone amounted to over $100,000. The total bill for the cleanup eventually reached more than $11 million.

> **"Estimates suggest that it takes the federal government approximately 72 hours to respond to a catastrophic disaster—a long wait for people who may be trapped or injured as a result of a natural disaster."**

Apart from the trained professionals and elected officials, local disaster response is often a grassroots operation. Town officials may work with church groups or accept donations of food, clothing, and money. Perhaps the biggest challenge in local disaster response is the tension between the federal and state governments and how this affects what goes on at the local level. Brett Water, a fire chief in Belgrade, Montana, sees relief as an issue of dollars, cents, and perspective: "Federal and state fire agencies are looking at what they can do to control costs," he says. "The local departments are just looking to keep fuel in their trucks."[29]

Private Organizations

Private relief organizations are a big part of disaster response in the United States. The American Red Cross is the most well known and the largest of these organizations. Each year, the American Red Cross responds to more than 70,000 disasters, setting up services in church halls, schools, and other public buildings. Four million people donate blood through the Red Cross, and much of it is used during medical emergencies related to disasters. With 35,000 employees and more than 500,000 volunteers, the symbol that gives this organization its name can be seen almost anytime disaster ravages a community.

Other groups, such as Feeding America, also work with victims of disasters. During Hurricanes Katrina and Rita, Feeding America and its sister organizations provided at least 65.2 million meals to victims of the two Gulf Coast storms. Like the Red Cross and so many other private organizations, Feeding America depends on donations.

In 2007 Priscilla Johnson and her friend Shuntrell Garrett learned firsthand about the generosity of Americans and the dedication of private relief organizations after fire forced them from their Dallas, Texas, apartment. "We had no idea how much they would help and we never thought we would need them. But there they were. They really helped."[30]

Estimates suggest that it takes the federal government approximately 72 hours to respond to a catastrophic disaster—a long wait for people who may be trapped or injured as a result of a natural disaster. Therefore, the responsibility for the immediate needs of those in crisis will likely remain at the state and local levels.

But under the most difficult circumstances, the federal government's role is an extremely important one. Congress and the president working with FEMA, can move essential resources such as airplanes and military personnel in a concerted and overarching way that neither states nor cities can.

Responses to recent disasters have shown that the system for dealing with large-scale tragedies is imperfect at best and sometimes incompetent. The question remains: How well will the United States handle its next natural disaster? Only time will tell.

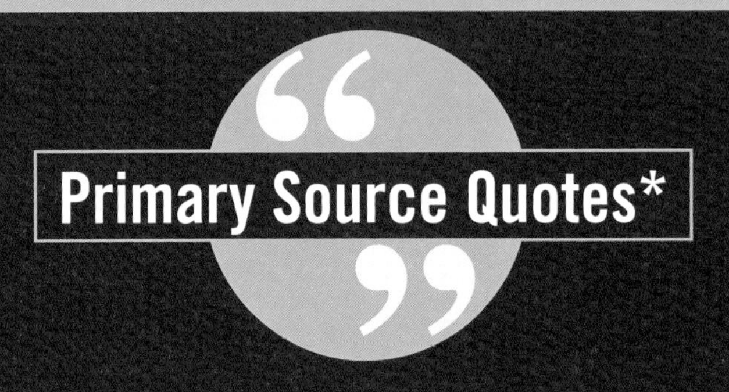

How Does the United States Respond to Natural Disasters?

"The slow, poorly coordinated and inadequate response to the New Orleans flood following Hurricane Katrina has raised serious questions on whether homeland security preparedness has really advanced the US's capacity to respond to natural disasters."

—Roger Few and Franziska Matthies, *Flood Hazards and Health*. London: Earthscan, 2006.

Few and Matthies are researchers at the University of East Anglia in Norwich, England.

"People said, well, the federal response was slow. Don't tell me the federal response was slow when there was 30,000 people pulled off roofs right after the storm passed."

—George W. Bush, "President Bush's Final News Conference (Transcript)," *New York Times*, January 12, 2009. www.nytimes.com.

Bush was the forty-third president of the United States.

Bracketed quotes indicate conflicting positions.

* Editor's Note: While the definition of a primary source can be narrowly or broadly defined, for the purposes of Compact Research, a primary source consists of: 1) results of original research presented by an organization or researcher; 2) eyewitness accounts of events, personal experience, or work experience; 3) first-person editorials offering pundits' opinions; 4) government officials presenting political plans and/or policies; 5) representatives of organizations presenting testimony or policy.

"It is very doubtful that even the best of administrations could sufficiently handle a direct hit upon Tampa or Miami, or a serious earthquake near San Francisco or Los Angeles. Our organizations are simply not up to preventing or handling large disasters."

—Charles Perrow, "What Is the State of U.S. Disaster-Preparedness?" *New York Times*, November 9, 2007. http://freakonomics.blogs.nytimes.com.

Perrow is a professor at Yale University.

"Three and a half years after Katrina there are still significant operational, communications and planning deficits in the Nation's emergency . . . system."

—U.S. Senate, Ad Hoc Subcommittee on Disaster Recovery, "Far from Home: Deficiencies in Federal Disaster Housing Assistance After Hurricanes Katrina and Rita and Recommendations for Improvement," February 2009. http://landrieu.senate.gov.

The U.S. Senate, Ad Hoc Subcommittee on Disaster Recovery is charged with coordinating long-term recovery operations, including coordination of data sharing between federal, state, and private agencies.

"We are on the right track to fulfilling our vision to become the nation's preeminent emergency management and preparedness agency."

—R. David Paulison, "National Hurricane Conference," April 4, 2007. www.fema.gov.

Paulison is a former director of FEMA.

"Katrina provided us with a worst-case scenario to see how fragmented and unprepared our entire health care delivery system is when disaster strikes."

—Chester W. Hartman and Gregory D. Squires, *There Is No Such Thing as a Natural Disaster*. Boca Raton, FL: CRC, 2006.

Hartman directs policy at the Poverty & Race Research Action Council; Squires is a sociology professor at George Washington University.

“The best way to ensure a strong and swift emergency response is to leave first response to first responders, leave decision-making in the hands of local and state leaders.”

—Rick Perry, “McCain Press Conference,” Office of Governor Rick Perry, April 3, 2006. http://governor.state.tx.us.

Perry is the current governor of Texas.

“The State of California must be prepared to plan and respond to earthquakes, fires, floods, disease or terrorist attacks. Right now, there is a good chance we would fall flat on our faces in a major crisis.”

—Pedro Nava, “State Disaster Response Needs Overhaul, Clear Chain of Command,” *California Progress Report*, May 3, 2006. www.californiaprogressreport.com.

Nava is a California assemblyman representing the state’s thirty-fifth district.

How Does the United States Respond to Natural Disasters?

- **1.1 million** people were displaced by Hurricanes Rita and Katrina in 2005.

- Funds raised to help the victims of Hurricanes Rita and Katrina amounted to **$33.8 million**.

- According to Feeding America, relief organizations provided at least **65.2 million meals** to victims of the two Gulf Coast storms.

- An estimated **83.5 million pounds** (38 million kg) of food and groceries were delivered to hungry victims in the aftermath of Hurricanes Katrina and Rita from 2004 to 2006.

- Each year, the American Red Cross responds immediately to more than **70,000 disasters**, including house or apartment fires (the majority of disaster responses), hurricanes, floods, earthquakes, tornadoes, hazardous materials spills, transportation accidents, explosions, and other natural and man-made disasters.

- Annually, tornadoes cause an average of **1,500 injuries** and roughly **80 deaths**.

Presidential Disaster Declarations

The following chart shows the number of disasters declared by the federal government, beginning with the administration of Dwight D. Eisenhower in the 1950s. Since then, the number of declared disasters for which the American government has provided financial support has grown from 15 under Eisenhower to 130 under George W. Bush.

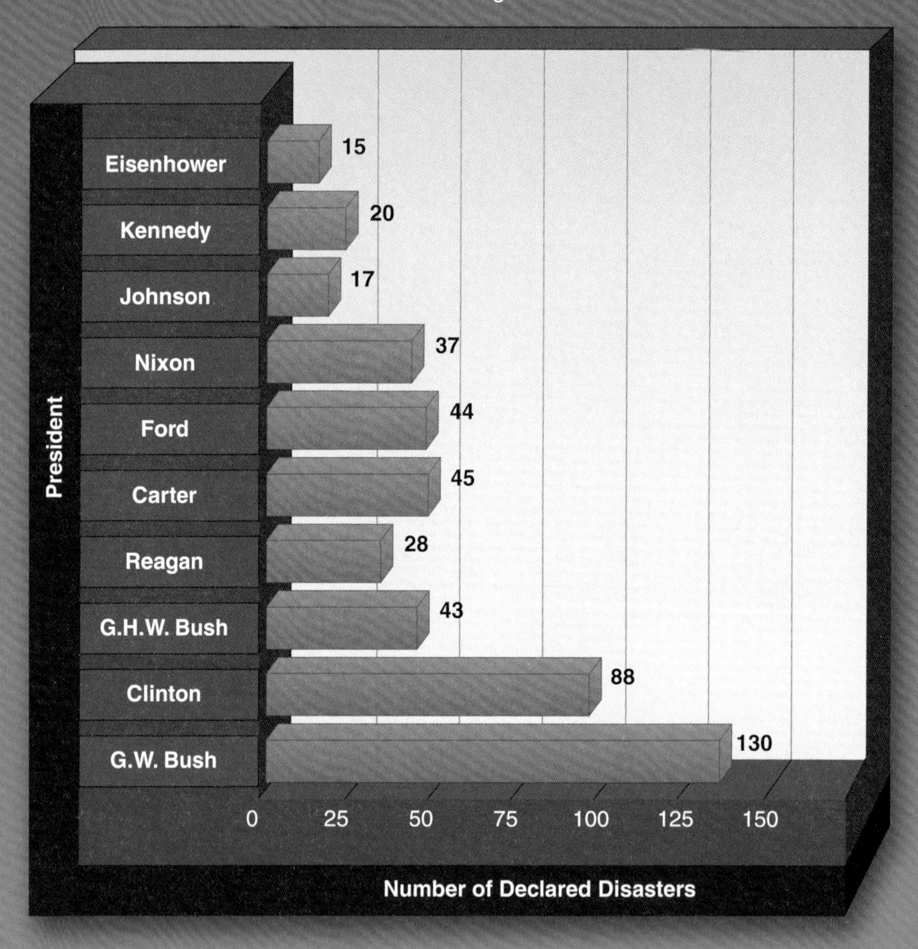

Sources: U.S. Department of Homeland Security, FEMA, "2008 Federal Disaster Declarations," April 7, 2008. www.fema.gov.

- In the United States, the most damaging 2008 event was Hurricane Ike, which struck Galveston, Texas, in September, killing 61 people and causing some **$30 billion** in damages along the Gulf Coast.

Americans Feel Disaster Response Lacking

A recent Pew Research poll reveals opinions on how various levels of government handled recent U.S. natural disasters. About half of survey participants approved of the state and local government response to midwestern floods in 2008 while a slightly higher percentage rated the federal government response as only fair or poor. Opinions on all levels of government response were more positive in connection with the 2007 California wildfires. Survey participants showed the most dissatisfaction with the government response at all levels to Hurricane Katrina in 2005.

Government's Response to:	State and Local Government	Federal Government
Midwestern floods (2008)		
Excellent/Good	51%	34%
Only Fair/Poor	35%	53%
Don't Know	14%	13%
California Wildfires (2007)		
Excellent/Good	76%	58%
Only Fair/Poor	16%	34%
Don't Know	8%	8%
Hurricane Katrina (2005)		
Excellent/Good	41%	38%
Only Fair/Poor	51%	58%
Don't Know	8%	4%

Source: Pew Research Center for the People & the Press, "Interest in Floods Increases, Still Lower than for '93 Deluge," June 25, 2008. http://pewresearch.org.

- The 2008 tornado season in the United States was severe, with about **1,700 tornadoes** causing several billion dollars in damages.
- **Flooding in the Midwest** during 2008 killed 24 people, injured 148, and forced 35,000–40,000 to evacuate their homes.
- Iowa's agricultural economic losses from floods exceeded **$2 billion** in 2008.
- In September 2008, Hurricane Gustav made landfall in Louisiana causing significant wind, storm surge, and flooding damage resulting in **$5 billion** worth of damage and **43** deaths.

2005 Was the Most Expensive Flood Loss Year Ever

2005, the year Hurricane Katrina hit the United States, was the most expensive year to date in terms of losses from flooding. During the second biggest flood loss year, 2008, the U.S. government estimates that it paid between $3 billion and $6 billion dollars in National Flood Insurance claims.

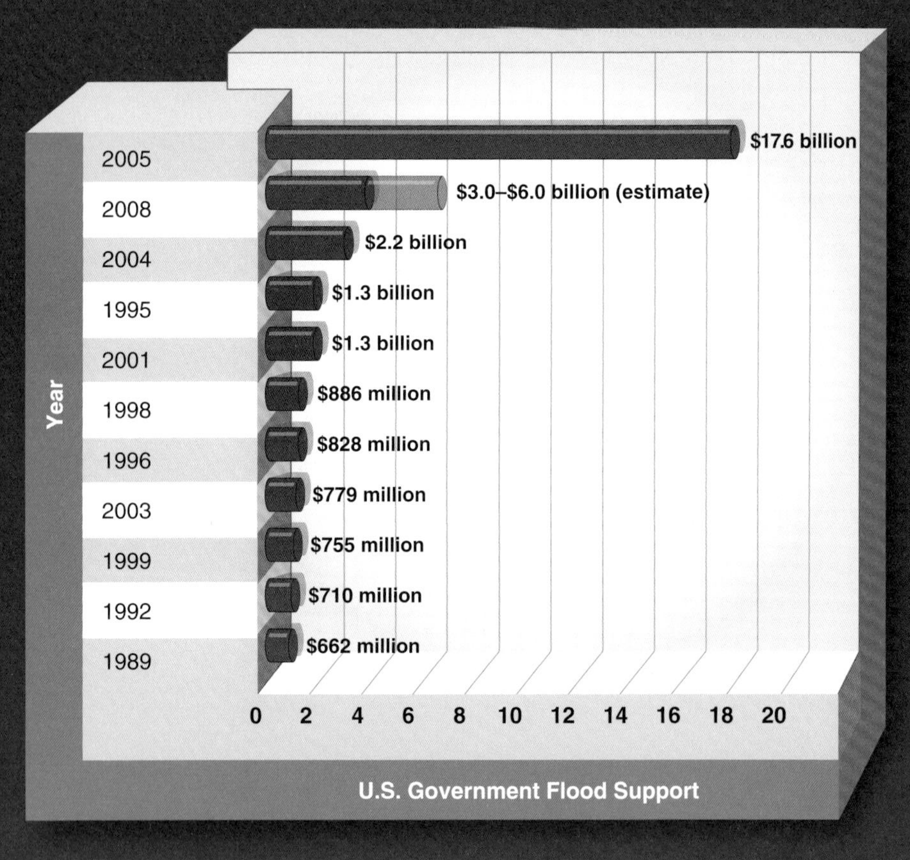

Source: Anthony R. Wood, "Hurricanes Prove Costly in 2008," *Philadelphia Inquirer*, November 27, 2008, p. A12.

- In summer and fall 2008, drought conditions across numerous western, central, and southeastern states result in thousands of wildfires; more than **5.2 million** acres burn, destroying more than 1,000 homes in California alone.

Is the United States Prepared for Man-Made Disasters?

"Every single day we collect more intelligence, share more information, inspect more baggage and passengers and containers, guard more territory and equip and train more first responders."

Tom Ridge, first U.S. secretary of homeland security.

"We looked at [the Pentagon's] plans. They're totally unacceptable. You couldn't move a Girl Scout unit with the kind of [nuclear, biological, and chemical attack] planning they're doing."

Arnold Punaro, retired U.S. Marine Corps major general, now serves as executive vice president at Science Applications International Corporation (SAIC).

According to a congressional commission report released in December 2008, the United States is in grave danger. The threats are not from earthquakes, volcano eruptions, or hurricanes. Experts say the most serious threat to the safety and stability of the country is a biological or nuclear attack. The report also suggests the Obama administration bolster its antiterrorism efforts by naming an official to oversee intelligence and foreign policy related to the spread of biological and chemical weapons. In this way, the United States can better prepare for a potential attack.

Biological Dangers

Five people died and dozens fell ill as a result of anthrax poisoning in the United States in October 2001. The victims included members of the media and congressional staff. The attack took place when envelopes containing a powder form of the anthrax spore arrived in the mail at offices around the country. Experts say the U.S. government seemed unprepared for such an attack. They warn that more such attacks are possible—or even likely—although next time the attacks may be even more dangerous. They worry that the next attack might involve smallpox. Commonly acknowledged as the most deadly virus ever to strike humankind, smallpox killed between 300 million and 500 million people worldwide between 1914 and 1977. Smallpox has no known cure; the major form of smallpox has a 30 percent mortality rate that can be reduced only through vaccination.

“Commonly acknowledged as the most deadly virus ever to strike humankind, smallpox has killed between 300 million and 500 million people worldwide.”

For government officials and scientists grappling with such biological threats, the nightmare scenario looks something like this: A terrorist group obtains a vial of smallpox, loads the virus into aerosol cans, and sprays it in a subway station in a major American city. In a matter of days, thousands, and soon millions, of people will likely be infected and die.

Preparing for the Worst

In 2003 the federal government committed $1.6 billion to state and local governments in an effort to prevent such a horrifying scenario. A portion of that money went into buying enough smallpox vaccine for every man, woman, and child in the country in the case of a biological attack.

In 2008, after a six-month investigation, the Commission on the Prevention of Weapons of Mass Destruction Proliferation and Terrorism concluded that preparation and response to a biological attack could be strengthened by a rapid medical response, better forensics to quickly identify the attacker, and continued research into better understanding the various viruses that pose the threats. "I think what we need to do is be imaginative but not hopeless," says Joseph Henderson of the Centers for Disease Control and Prevention (CDC), "not to be morbid but to think through how in fact we will survive the event."[31]

The Nuclear Threat

Officials worry that nuclear weapons, or the material used to make them, could fall into the hands of terrorists or enemy nations. Today, 32 countries harbor 3,200 tons (2,903 metric tons) of fissile material—matter such as uranium or plutonium needed to set off a nuclear chain reaction. This is enough to make 240,000 nuclear weapons.

And while all-out nuclear war remains unlikely, officials do fear the use of a "dirty bomb." The "dirty bomb" idea, in which a terrorist could set off a crude but highly destructive device that spreads radioactive material packed into a bag or suitcase, is difficult to plan for. The U.S. Nuclear Regulatory Commission (NRC) works with the Department of Homeland Security and FEMA to develop plans for preventing and responding to a nuclear attack. In recent years it has strengthened its nuclear licensing programs, making it more difficult for those seeking fissile material to get it.

Yet, in 2007 investigators from the Government Accountability Office (GAO) tested the commission's system and found it lacking. GAO investigators created a fake company and, after filling out the required paperwork, easily obtained a license from the NRC to buy the radioactive materials needed to build a "dirty bomb." The NRC did not investigate the false company, nor did it visit the company's offices. Members of Congress rely on these kinds of tests to reveal holes in disaster preparedness. In their report, officials warned, "The terrorists have been active, too, and in our judgment America's margin of safety is shrinking, not growing."[32]

Preparation for Nuclear Attack

Today, the three greatest nuclear threats to the United States are Pakistan, Iran, and North Korea. American officials are most concerned about Pakistan, an unstable country with nuclear weapons. Terrorist groups within Pakistan include al Qaeda and its ally the Taliban, both of which partly control the country's western border area. If Pakistan's shaky government collapses, say terrorism experts, the United States will face a crisis: Terrorists could launch a nuclear attack in the region or smuggle the weapons out of the country and threaten the United States.

Experts say that the United States is unprepared for a nuclear attack of any size. In a study published in 2007, the University of Georgia's Center for Mass Destruction Defense reported that if a nuclear device was set off in a major city such as Chicago, Atlanta, New York, or Washington, D.C., the hospitals needed to treat the estimated 5 million injured would likely be destroyed in the blast. Also, the supply of mobile hospital beds to replace the hospitals would be insufficient.

> **"Today, the three greatest nuclear threats to the United States are Pakistan, Iran, and North Korea."**

The study, conducted over three years, also sets priorities for preparing for and responding to such an event. The study states that "accelerated training and coordination among the federal agencies tasked for WMD [weapons of mass destruction] response, military resources, academic institutions, and local responders will be critical for large-scale WMD events involving mass casualties."[33] Scholar Graham Allison says the most important thing the federal government can do during a biological or nuclear attack "is to educate and communicate with the public . . . authorities can provide the media with a consistent instructive message for the frightened public."[34] This, he believes, could minimize the catastrophe by keeping people calm and getting them the help they need quickly.

Crumbling Infrastructure

A fatal bridge collapse in Minneapolis in 2007 shocked many Americans as they watched the tragedy unfold on television. Investigators concluded that the collapse of the nearly 50-year-old bridge was due to faulty additions to the structure such as concrete lane barriers. One investigator, after being asked whether such flaws were typical of America's bridges, said he couldn't be sure: "This could well be a one-off thing. But you don't know that."[35]

The American Society of Civil Engineers (ASCE) recently gave the country's infrastructure a "D" grade. Presently, they say, at least 160,570

American bridges suffer from serious defects, 33 percent of roads are in poor to mediocre condition, and 4,000 dams are deficient and could fail. Such sobering statistics worry state and local officials and have many Americans wondering whether the nation is prepared for the potential disasters that aging and disrepaired infrastructure might bring.

> **"The Environmental Protection Agency, a government organization in charge of protecting the nation's natural habitats, has a disaster response plan in place to get trained personnel to the site of an oil spill within 24 hours."**

In 2009 interest in rebuilding some of the nation's infrastructure revived when the U.S. Congress passed a massive stimulus bill to revive the country's flagging economy. According to the plan, billions of dollars would be pumped into state economies across the nation for the express purpose of securing the country's infrastructure. But critics complained the allocated money would account for only 5 percent of the $2.2 trillion required to do the job and that what money there was would not be spent fast enough to prevent more disasters: "The main problem with infrastructure is the timetable," says commentator Mortimer B. Zuckerman. "Only 40 percent of the proposed program would be spent in the first two fiscal years."[36]

Most state and local governments cannot wait that long. In many cities, the water and sewage systems are 100 years old. Built in the early twentieth century, New York City's 2 main underground tunnels pump 1.3 billion gallons (4.9 billion L) of water to residents daily. Both leak badly and are barely functional. A third, more modern tunnel has been under construction since 1969 but will not be completed until 2020. Political officials and tunnel workers fear the city is unprepared for the potential disaster and may not understand the seriousness of the situation. "If one of those tunnels goes, this city will be completely shut down," says Jimmy Ryan, a longtime tunnel worker. "In some places there won't be water for anything. Hospitals. Drinking. Fires. It would make September 11 look like nothing."[37]

Oil Spills

Oil spills in the United States are a fairly common occurrence. Although exact figures are difficult to ascertain, hundreds happen each year. Many of these spills are small and contained and will be cleaned up in a matter of hours. Large spills, on the other hand, can take weeks or even months to bring under control. The Environmental Protection Agency, a government organization in charge of protecting the nation's natural habitats, has a disaster response plan in place to get trained personnel to the site of an oil spill within 24 hours. Once on the scene, workers use chemicals and gels to contain and dissolve the black sludge in hope of reducing damage to habitats and wildlife.

> **"By now, passengers were lining up on the airplane's wings, waiting. Fighting the river currents, the rescue boat pilots moved in as closely as possible until victims were able to jump aboard or be carried to safety by police officers or other emergency workers."**

A November 2007 spill dumped 58,000 gallons (219,489L) of oil into San Francisco Bay. Crews quickly began work on preventing the oil from spreading; they used foam-filled, fencelike booms, usually made of urethane or PVC, to keep the sludge in one place. Despite a volunteer crew of 400 people, in the days after the spill some of the oil sank to the bottom of the bay or drifted away in thin layers. The continued cleanup cost the city millions of dollars and countless resources.

Transportation Disasters

The most common man-made disasters are those involving planes, trains, and automobiles. Over 31,000 people died on America's highways in 2008, mostly in accidents caused by human error. The National Transportation Safety Board handles the federal response to transportation disasters, but typically such accidents are first addressed at the local level. Each day, first responders are called to the scene of highway crashes to tend to the injured.

And while plane and train crashes are less common than those in cars, they are more complicated to deal with mainly because they are so much larger and heavier and because so many more people are involved. In the aftermath of a Los Angeles train crash that killed 25 and injured 135 in the fall of 2008, firefighters enlisted over 30 fire trucks, dozens of ambulances, and more than 600 feet (183m) of fire hose to aid the injured passengers. Triage centers were set up at the scene to provide immediate medical care.

When a US Airways flight crash-landed in the Hudson River in January 2009, the plane was in danger of sinking, trapping 155 passengers. The winter temperatures also worried Coast Guard and river rescue units: People could freeze to death after only a few minutes in the frigid water. Preparation and rapid response would be crucial in saving lives.

In moments, helicopters and high-speed boats were on the scene of the crash. Divers in scuba gear moved close to the aircraft. By now, passengers were lining up on the airplane's wings, waiting. Fighting the river currents, the rescue boat pilots moved in as closely as possible until victims were able to jump aboard or be carried to safety by police officers or other emergency workers. New York firefighter Tom Sullivan attributes the successful response to the organized and calm atmosphere created by his fellow responders. "It's unbelievable," he says. "This is a large airliner with this many people, it worked out well. I was amazed that no one was killed."[38]

Preparation, Sullivan believes, enabled his unit to do its job quickly and effectively. But luck is also an essential component to disaster response. The pilot landed the plane smoothly, the plane did not sink, and no one was trapped inside. These factors only provided an opportunity for success. It was up to the rescue workers to seize that opportunity and save lives.

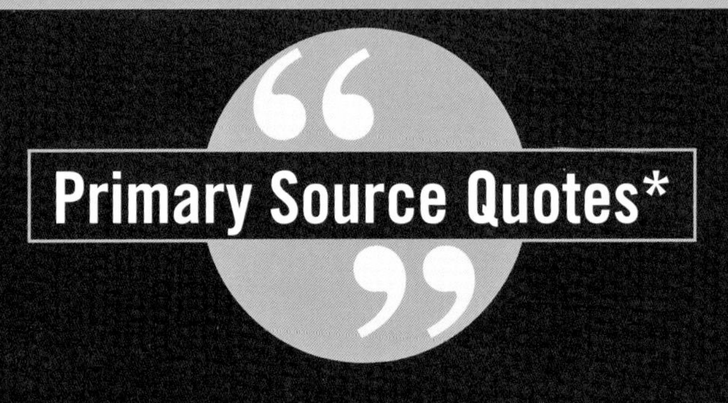

Is the United States Prepared for Man-Made Disasters?

"The United States still does not have and must quickly develop a fully comprehensive and tested system for the rapid delivery of lifesaving medical countermeasures against anthrax and other bioterrorist threats."

—United States Congressional Commission on the Prevention of Weapons of Mass Destruction Proliferation and Terrorism, *Work at Risk: Biological Proliferation and Terrorism*, 2008. www.preventwmd.gov.

The bipartisan commission made recommendations on how to better prevent and prepare for nuclear, biological, or chemical terrorism.

"We are relatively well prepared against biological attacks, and their impact would likely be limited."

—Thomas McInerney, *Endgame*. Washington, DC: Regnery, 2005.

McInerney is a retired U.S. Air Force lieutenant general and a contributor to Fox News.

* Editor's Note: While the definition of a primary source can be narrowly or broadly defined, for the purposes of Compact Research, a primary source consists of: 1) results of original research presented by an organization or researcher; 2) eyewitness accounts of events, personal experience, or work experience; 3) first-person editorials offering pundits' opinions; 4) government officials presenting political plans and/or policies; 5) representatives of organizations presenting testimony or policy.

“Most of us, even those with military experience, have never encountered patients exposed to these [biological or chemical] weapons and are ill-prepared to treat them.”

—M. Gage Ochsner Jr., “Commentary,” in *Trauma*, by David V. Feliciano, Kenneth L. Mattox, and Ernest E. Moore. New York: McGraw-Hill Professional, 2007.

Ochsner is a physician practicing in Georgia.

“The past eight years have seen almost every nuclear proliferation problem grow more dangerous. U.S. policy has not only failed to reduce these dangers, in many cases it made them worse.”

—Joseph Cirincione, “The New US Policy: Securing the World from Nuclear Threats,” speech to European Parliament, Brussels, December 9, 2008. www.wiersma.pvda.nl.

Cirincione is president of the Ploughshares Fund and an expert on nuclear proliferation.

“Terrorists are more likely to be able to obtain and use a biological weapon than a nuclear weapon.”

—Commission on the Prevention of Weapons of Mass Destruction Proliferation and Terrorism, “Report: Executive Summary,” December 2, 2008. www.preventwmd.gov.

The bipartisan commission was led by Senators Bob Graham and Jim Talent.

“In the event of a biological attack, public health officials may not immediately be able to provide information on what you should do.”

—Jane A. Bullock, George D. Haddow, and Damon P. Coppola, *Introduction to Homeland Security*. St. Louis, MO: Butterworth-Heinemann, 2006.

Bullock, Haddow, and Coppola are experts in the field of disaster response.

“The risk of nuclear terrorism may be overstated.”

—Nilay Cabuk Kaya, Aykan Erdemir, *Social Dynamics of Global Terrorism and Prevention Policies*. Amsterdam: IOS, 2008.

Kaya and Erdemir are experts on global terrorism.

“The engineering society of the country gave us [the United States] a D grade on infrastructure. . . . We’ve got to get serious about this. We’ve got to fix America’s dams, tunnels, bridges, highway systems.”

—Barry McCaffrey, “Minnesota Bridge Collapse,” *Hardball*, June 28, 2007. http://video.msn.com.

McCaffrey is a retired U.S. Army general.

Is the United States Prepared for Man-Made Disasters?

- According to a 2006 study by the Centers for Disease Control and Prevention, **88 percent** of American hospitals say their nurses have training in recognizing and treating exposure to at least one of the diseases associated with bioterrorism (smallpox, anthrax, plague, botulism, tularemia, viral encephalitis, and hemorrhagic fever).

- After the 2001 attacks, the U.S. government stockpiled more than **300 million doses of smallpox vaccine**—enough for every man, woman, and child in the United States.

- According to a 2008 survey of national security preparedness by the Radiological Threat Awareness Coalition
  - **81 percent** of Americans consider the threat posed to the United States by the possibility of a "dirty bomb" attack by terrorists to be serious.
  - **56 percent** of Americans polled said they were not confident they would know what to do in the event of a "dirty bomb" attack.
  - **63 percent** said they do not feel prepared in the event of a terrorist attack on their community.
  - **34 percent** of Americans believed the government was doing a good job informing people about preparedness for a terrorist attack.

Bioterrorism Threats

This map of the United States shows the relative bioterrorism threat levels that exist from region to region. The areas in red signify the cities or towns where the danger of bioterrorist attacks is highest. Medium threat levels are in purple, and minimal threat levels are in blue. Such visuals help federal and state officials decide where disaster response resources could best be used.

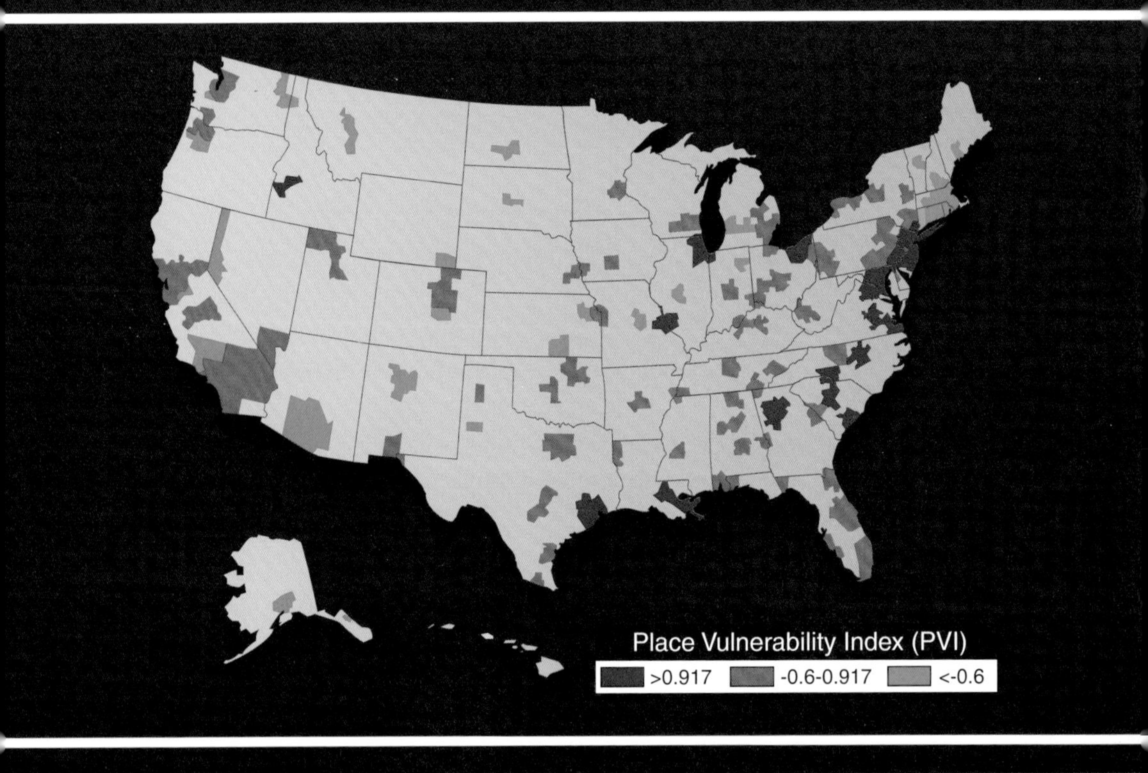

Source: *Spatial Sustain,* "Bioterrorism Threat Mapped for U.S. Cities," March, 2008. http://vector1media.com.

- In hospitals that had 24-hour emergency departments or outpatient clinics, **86 percent** of clinical staff members are trained to recognize and treat smallpox, and **82 percent** are trained to recognize and treat anthrax infections.

Smallpox Is a Threat

Smallpox, a horrific virus that American leaders fear could be unleashed on the American public by terrorists, is highly contagious. As this graphic illustrates, variola, the virus that causes smallpox, is spread through saliva droplets in an infected person's breath. Its incubation period is brief, and it has the potential to kill millions of people. While American health officials have worked to stockpile vaccines in case of an attack, the release of smallpox could cause widespread panic, making response efforts more difficult.

What is smallpox?

- Airborne virus, highly contagious
- Caused by variola virus
- Virus related to monkeypox
- Could take approximately 6 weeks to seed smallpox cases around the world
- Each infected individual infects an average of 20 others

Early symptoms

- High fever, fatigue, and rash
- Resulting spots fill with clear fluid and pus then form a crust and fall off
- Fatality rate ranges from 15 to 50 percent

Treatment

- Vaccination (vaccinia) to prevent infection
- During the first 4 days after exposure
- Current U.S. vaccine stockpile contains 300 million doses
- To be kept on shelves and used in confirmed cases only

History

- **1976** WHO urges labs worldwide to destroy retained stocks of variola virus
- **1977** Last known natural case—Somalia
- **1978** Laboratory accident in Birmingham, England, kills one and causes a limited outbreak
- **1983** South Africa—last country to destroy its virus stock

Effects on human body

- Spreads through saliva droplets in an infected person's breath

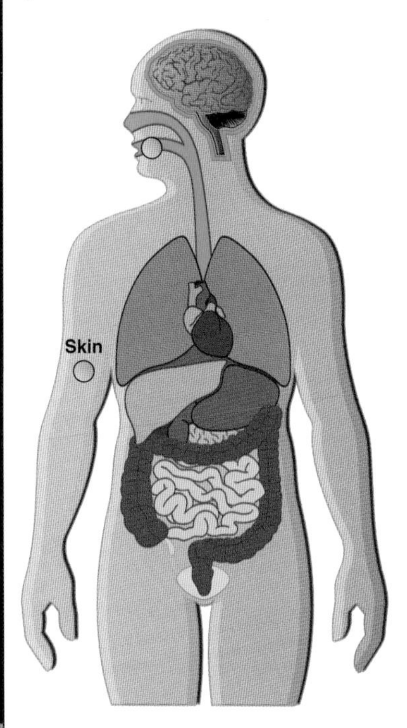

- Incubation period of 7–17 days after exposure
- Infectious once rash develops and first week of illness

Source: WearCam/Reuters, "Smallpox in Our Midst/Smallox Threat," December 11, 2002. http://wearcam.org.

Nuclear Danger Zones

The maps below illustrate the current state of nuclear weapons programs around the world as well as changes since the 1960s. In the top map, the countries in purple—the United States, Great Britain, China, and Russia—have signed nonproliferation treaties. The goal is to keep nuclear weapons out of the hands of terrorists or unstable nations. In yellow are countries with suspected weapons programs.

Countries With Nuclear Weapons or Programs

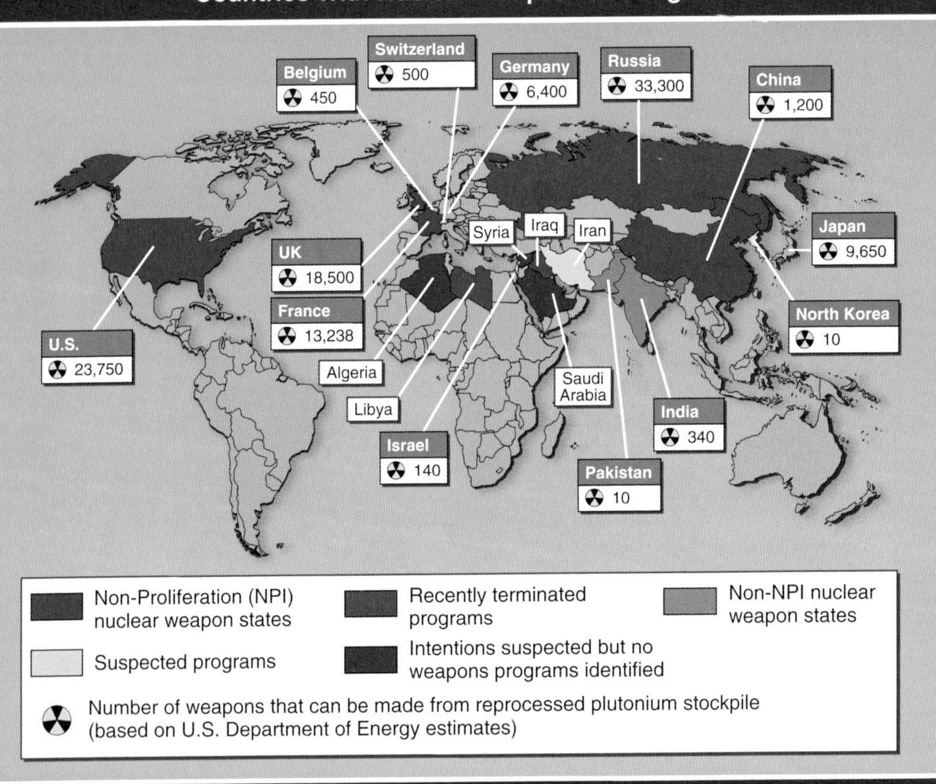

Non-Proliferation (NPI) nuclear weapon states
Recently terminated programs
Non-NPI nuclear weapon states
Suspected programs
Intentions suspected but no weapons programs identified
Number of weapons that can be made from reprocessed plutonium stockpile (based on U.S. Department of Energy estimates)

Nuclear Weapons Programs Declining

Source: *New Scientist*, "The Spread of Nuclear Weapons," 2004.

Water Infrastructure at Risk

The Environmental Protection Agency created this projection illustrating the need for improved water infrastructure. The sharp rise suggests a potentially dangerous future. Experts fear aging pipes could break down or terrorists could exploit weaknesses unless outdated water systems are fixed or replaced.

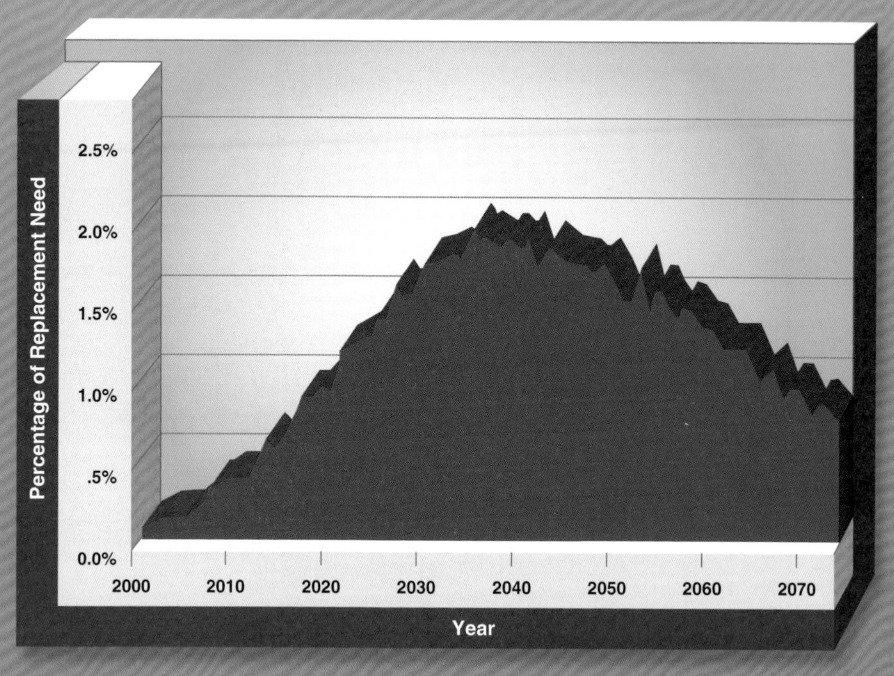

Source: *The National Academies' Water Information Center*, "Who Is in Charge of America's Taps? The Regulatory Framework," 2008. http://water.nationalacademies.org.

- Greenpeace International estimates that approximately **30,000 nuclear weapons** are held globally.

How Can Disaster Response Be Improved?

“Promoting an emergency and trauma care center that works well . . . is fundamental to establishing a system that will work well in the event of a disaster.”

—The Institute of Medicine, which is charged with assessing and improving the quality of health and disaster care in the United States.

“Effective disaster management requires a holistic approach that integrates preparedness, response, recovery, and risk reduction.”

—Susan Johnson, director of the National Societies and Field Support Division, part of the United Nations General Assembly.

The Chinese Earthquake Response

The deadliest earthquake to strike China in 30 years occurred on May 12, 2008. The 7.8-magnitude quake, with its epicenter in Sichuan Province, killed approximately 70,000 people and injured about 375,000 others. At least 900 students were buried under the rubble of a secondary school in the city of Dujiangyan; bystanders struggled for hours to free victims pinned under fallen beams and slabs of concrete. In Shifang City, to the southeast, a chemical plant collapsed, trapping hundreds of people and sending tons of liquid ammonia, a toxic substance, into the environment.

Response efforts were complicated because Sichuan is one of the poorest areas in the country. "There are a lot of people living in marginal areas that are difficult to get to," said Dale Rutstein of UNICEF China, "and a lot of the buildings in those areas are substandard and could collapse very quickly."[39]

Chinese authorities immediately ordered a massive rescue effort, which initially included 50,000 soldiers and police and nearly 200 emergency relief workers. In short order, helicopters also began delivering food, water, and medical supplies to the hardest hit areas. Prime minister Wen Jiabao released a statement calling for calm, confidence, and courage; state television broadcast reports advising victims that food, water, and help was on the way.

> **Chinese authorities immediately ordered a massive rescue effort, which initially included 50,000 soldiers and police and nearly 200 emergency relief workers.**

But these rescue efforts were only the beginning: In the days that followed, thousands more troops and militia members joined in the relief operations. The day after the earthquake, the Chinese government also accepted aid from Taiwan's Tzu Chi Foundation, the first outside help allowed. Rescue organizations from Japan, Singapore, Russia, South Korea, and the United States soon followed.

Despite the heavy toll, relief efforts were widely praised. Response was rapid, organized, and saved hundreds of thousands of lives. Chinese vice premier Hui Liangyu later spoke about the importance of responding effectively to disasters: "Improving emergency response and disaster relief is an important task in building a harmonious society,"[40] he said.

Developing an Effective Response to Disasters

Following the 2004 tsunami, public health experts feared thousands of additional deaths from disease and hunger. These fears did not materialize. "Instead, the relief effort was effective as food, medicine, and other goods were delivered as fast as transportation could be arranged," according to a 2005 analysis. "The survey reflected the excellent logistics work accomplished, perhaps resulting from the strong financial response to this disaster."[41]

That the response to such an enormous disaster received such a positive assessment can be seen as a significant accomplishment. However, the same analysis also noted several areas in need of improvement. One of these involved collaboration and coordination between disaster response agencies. So many agencies responded to the immense needs of the region that, in some cases, they ended up competing with each other over who would provide assistance, what would be provided, and who it would go to. Also, according to the analysis, coordination between agencies met immediate needs of people in the region but fell short in meeting more long-term needs.

Strong Leadership Needed

Improving disaster response requires strong leadership, according to most experts in the field. But strong leadership is lacking in many countries. Great Britain's international development minister Gareth Thomas has called for more properly trained coordinators to oversee disaster relief efforts around the world. According to Thomas, 33 percent of the countries most in danger of experiencing a humanitarian emergency do not have people who can lead relief efforts. The lack of trained and experienced people reduces the likelihood of providing a coordinated and efficient response. Thomas believes organizations need to step up efforts to find and train leaders in the countries that are most vulnerable to disasters. He says, "We need humanitarian agencies, the Red Cross and NGOs to put forward candidates; we need more candidates from developing countries and more women."[42]

> **A 2006 White House report made 125 recommendations for improved disaster relief, including better plans for evacuating victims, tracking supplies and delivering quick information from crisis zones.**

Developing leadership includes training local people who can step in when the need arises. Having a common language and familiarity with local culture can greatly improve the supply chain—making sure that the right supplies reach people in need. Having more trained female relief workers is also essential as victims of rape during wartime may be more comfortable with female doctors, nurses, and caregivers.

Better Tools for Communication

One essential component for getting disaster victims the aid they need is improved communication. "When disaster strikes," says U.S. commerce secretary Carlos M. Gutierrez, "first responders must have the tools to communicate."[43] Recent global efforts to secure these tools have been beneficial.

Political elections in Kenya in 2007 sparked violence, forced many people from their homes, and left thousands without food and water and in need of medical care—some for extended periods of time. As a way of communicating with those victims and getting food and medicine to where it was most useful, a coalition of Kenyan NGOs working with Oxfam created PeaceNet. The idea was to create a text-messaging "nerve center" by providing real-time information about potential violence. PeaceNet received hundreds of useful warnings from civilians. One read, "We have been alerted that it is not safe tonight, in Bamburi, Utange, home area. We are asking 4 security here please." Another told aid workers that "the situation in Narok south is bad. People have camped at the catholic church in Mulot and there are fears that they may be attacked tonight."[44] With the information literally in hand, local "peace committees" were sent out to try to calm things down and reach those in need of food and medical care.

Fixing FEMA and Other Recommendations

In the United States, a 2006 White House report made 125 recommendations for improved disaster relief, including better plans for evacuating victims, tracking supplies, and delivering quick information from crisis zones. The report also called for a new and improved National Response Plan to serve as a blueprint for handling large-scale calamities, and closer collaboration with the military: "The departments of Homeland Security and Defense should jointly plan for the Department of Defense's support of federal response activities," reads the report, "as well as those extraordinary circumstances when it is appropriate for the Department of Defense to lead the federal response."[45]

Also in 2006, a Senate report offered an additional 86 recommendations, most notably replacing FEMA with a stronger successor that can more quickly respond to catastrophe. The new organization, which the report named National Preparedness and Response Authority, would serve as the president's primary adviser on disaster management. Although seriously considered by government officials, FEMA, thus far, remains intact.

"A primary goal of fixing FEMA is to restore public confidence by assuring Americans that when overwhelming catastrophe strikes, the federal government will be on hand to help."

FEMA's budget expanded to $8 billion in 2008, an 11 percent increase over the previous year. New initiatives included hiring more staff, paying for new technologies, and increasing rescue supplies for states most vulnerable to natural or man-made disasters. Officials were determined to put the agency on a solid path to more effective disaster response and not to repeat past mistakes.

For officials, a primary goal of fixing FEMA is to restore public confidence by assuring Americans that when overwhelming catastrophe strikes, the federal government will be on hand to help. Only time will tell whether political officials and FEMA itself will be able to reorganize its priorities and restore its damaged reputation.

Preparing for the Worst

According to national defense expert Lawrence P. Farrell Jr., preparation is essential in responding to large-scale disaster. He mentions that a common management saying is that "plans are only good intentions unless they immediately translate into effective action."[46] Experts say that in a major disaster, emergency workers may not be able to reach everyone right away, and in some cases it may take three or more days for help to arrive. They encourage families to devise a plan to help ensure their safety and comfort during a disaster. "You have to feel like you are an agent in your own survival," says disaster reporter Amanda Ripley. "You and your co-workers and neighbors are going to be there, not homeland security paratroopers. The more confidence you have before the event happens, the less debilitating the fear will be and the better your performance will be."[47]

A unique experiment took place in November 2008. The Earthquake Country Alliance of San Diego County, California, organized the Great Southern California ShakeOut, a massive simulation of what might happen if a major earthquake struck the area. Participating in what was advertised as the largest earthquake drill in history were 400,000 people, including 325,000 children in 43 school districts.

"Video, wireless, and handheld devices will enable workers to integrate the various pieces of the response puzzle and communicate in a more seamless, real-time way than they are currently able to do."

When the alarm bell rang in schools across the area, students dropped to the floor or crouched under desks. Bishop Alemany High School students wore fake blood to make things more realistic, and local medical centers rushed to the scene, testing their emergency systems in an effort to someday save lives. Local hospitals used the drill to better prepare emergency rooms and organize their trauma centers in ways that would save the most lives during a real disaster. According to California governor Arnold Schwarzenegger, that day was a total success: "The locals, the state and federal government came together very quickly," he said, "unlike what we have seen at [Hurricane] Katrina, when it was going the other way."[48]

Hospitals and Emergency Teams

A 2007 poll reports that 6 in 10 Americans say they will be ready when a natural disaster strikes their community. A majority also believes that local hospitals and emergency teams are prepared to help in case of danger. That may not be the case, suggests Georges Benjamin, executive director of the American Public Health Association (APHA). "For too long," says Benjamin, "public health and medicine have responded to emergencies in separate silos."[49] In other words, the different parts of the system do not always know what the other parts are doing. This leads to repetition, miscommunication, and the sending of medical supplies to hospitals and centers that may not need them.

In 2007 the APHA released a list of 53 recommendations on how to improve medical care in response to a disaster. They called on more funding from the federal government; more training and education for public health professionals; and better integration of emergency systems between federal, state, and local authorities. According to Judith Woodhall, executive director of the medical group Comcare, "When emergency responders arrive at an emergency scene, the victim or the patient is really a blank slate to them. So we're looking to create informed emergency responders throughout the continuum of care during an emergency event."[50]

The Role of Technology

According to General Victor Renuart, commander of the North American Aerospace Defense Command and U.S. Northern Command, a team's ability to oversee a major disaster relief effort will be essential. "There will be a time when the size of the event is so big [and] happens so quickly," he says, "that you have to have an integrated team of local and state and federal responders, both from the military and from our civilian first responders."[51] Video, wireless, and handheld devices will enable workers to integrate the various pieces of the response puzzle and communicate in a more seamless, real-time way than they are currently able to do.

A recent $4 million project funded by the National Library of Medicine attempts to do just that. One of its new technologies uses wireless to help first responders track a disaster victim's location and health status; another provides 3-D visuals for use by emergency workers.

One project organizer emphasized the importance of such technology in meeting disasters head-on: "Law enforcement is an integral part of medical disaster response," he said, "and to better coordinate that, they anticipate that technologies like this can be useful in communicating from law enforcement to medical responders."[52]

Experts in the field of disaster response continue developing new ways to cope with crises around the world. And while natural and man-made disasters may be inevitable in the twenty-first century, the skill, diligence, and resources required to handle them can always be improved.

How Can Disaster Response Be Improved?

"We will improve information sharing, strengthen our enforcement mechanisms and intensify accountability, and we will provide more effective means for the private sector to join us in meeting our goals for the safety and security of our nation."

—Janet Napolitano, "Confirmation Hearing Statement," *Essential Documents*, January 15, 2009. www.cfr.org.

Napolitano, former governor of Arizona, is now secretary of the Department of Homeland Security.

"FEMA will never prepare for catastrophic disasters if it continues to spend its finite time, money, resources, and personnel on every disaster that happens in America."

—Jena Baker McNeill, "Key Questions for Janet Napolitano, Nominee for Secretary of Homeland Security," Heritage Foundation, January 13, 2009. www.heritage.org.

McNeill is a homeland security policy expert.

* Editor's Note: While the definition of a primary source can be narrowly or broadly defined, for the purposes of Compact Research, a primary source consists of: 1) results of original research presented by an organization or researcher; 2) eyewitness accounts of events, personal experience, or work experience; 3) first-person editorials offering pundits' opinions; 4) government officials presenting political plans and/or policies; 5) representatives of organizations presenting testimony or policy.

“Almost every day brings reports of serious damage and loss of life during a storm, flood, drought or other natural hazard. Without concerted action, we could see natural catastrophes on an unprecedented scale.”

—Ban Ki-moon, “Remarks to the Ministerial Meeting on Reducing Disaster Risks in a Changing Climate,” UN News Centre, September 29, 2008. www.un.org.

Ban Ki-moon is secretary general of the United Nations.

“During disaster responses, the actions of individuals are crucial to rescue lives and to avoid further losses.”

—Claudia Seifert, in *Communicable Crises*, Deborah E. Gibbons, ed. Charlotte, NC: Information Age, 2007.

Seifert is an expert in disaster response.

“A post-disaster analysis of the entire response should be carried out within 24 to 48 hours of its completion, to identify weaknesses and the lessons that should be learned to improve future responses.”

—Lewis Flint et al., *Trauma*. Hagerstown, MD: Lippincott, 2007.

Flint is a trauma surgeon.

“Drills serve a valuable purpose in testing the adequacy of a disaster plan, as well as [emergency] personnel’s knowledge of the plan.”

—Jerrold B Leikin and Robin B. McFee, *Handbook of Nuclear, Biological, and Chemical Agent Exposures*. Evanston, IL: Informa HealthCare, 2007.

Leikin is an expert in emergency medicine; McFee is a professor in the Department of Preventive Medicine at SUNY–Stonybrook.

“Every community should have a disaster plan. This requires coordination and collaboration among myriad agencies and disciplines.”

—American Medical Association/American Public Health Association, “Improving Health System Preparedness for Terrorism and Mass Casualty Events: Recommendations for Action,” July 2007.

The AMA is the largest association of doctors and medical students in the United States. The APHA is the oldest medical organization in the world.

“Experience has shown that good planning can greatly reduce technical problems . . . that can otherwise seriously undermine the speed and effectiveness of aid provided to stricken communities.”

—Michael Schulz, “Coordination of Humanitarian and Disaster Relief Assistance,” in *International Federation of Red Cross and Red Crescent Societies: Speeches and Statements*, November 2008. www.ifrc.org.

Schulz is a permanent observer to the United Nations.

How Can Disaster Response Be Improved?

- In March 2008 the National Oceanic and Atmospheric Administration (NOAA) deployed the final two **tsunami detection buoys** in the South Pacific, completing the buoy network and bolstering the U.S. tsunami warning system.

- Determination of earthquake location and size in Indonesia dramatically improved in 2007 due to the **German Tsunami Early Warning Project**, cutting warning time from 17 minutes to 2 minutes and 11 seconds.

- In 2008 the U.S. Army purchased **mobile command vehicles**. The vehicles include a system called Sentinel, which will provide for better communication during disasters.

- A survey of registered nurses in Missouri found that most had not received any education in **bioterrorism**, putting the state's citizens at risk.

- The U.S. Forest Service currently spends **48 percent** of its budget fighting forest fires.

- **Disaster-modeling software**, inspired by video games, is helping responders create rescue plans in real time.

Improving FEMA

In April 2008 the Department of Homeland Security released its report card on FEMA's efforts to improve its response to disasters in the United States. The agency made substantial progress in its evacuation readiness, yet its progress in coordinating and supporting a response was deemed only modest.

FEMA Report Card

The Department of Homeland Security inspector general rated the Federal Emergency Management Agency's progress in improving its disaster-response preparedness in nine categories on a scale ranging from limited (worst) to substantial (best):

Category	Level of Progress
Overall planning	Moderate
Coordination and support	Modest
Interoperable communications	Moderate
Logistics	Moderate
Evacuations	Substantial
Housing	Modest
Disaster workforce	Modest
Mission assignments	Limited
Acquisition management	Moderate

Key to Scale

Limited or no progress:
There is an awareness of the critical issues needing to be addressed, but specific corrective actions have not been identified.

Modest progress:
Corrective actions have been identified, but implementation is not yet under way.

Moderate progress:
Implementation of corrective actions is under way, but few if any have been completed.

Substantial progress:
Most or all of the corrective actions have been implemented.

Sources: FEMA Inspector General's Office, "FEMA Report Card," 2008; Bill Walsh, "FEMA Better Prepared, Report Says," *New Orleans Times-Picayune*, April 3, 2008. www.nola.com.

Wildfire Snapshots Useful to Firefighters

The map below illustrates California wildfire activity on July 13, 2008, at 7:00 A.M. This snapshot in time was used to immediately update firefighters and aid workers as they tried to contain and put out the blazes in more than 20 fire zones.

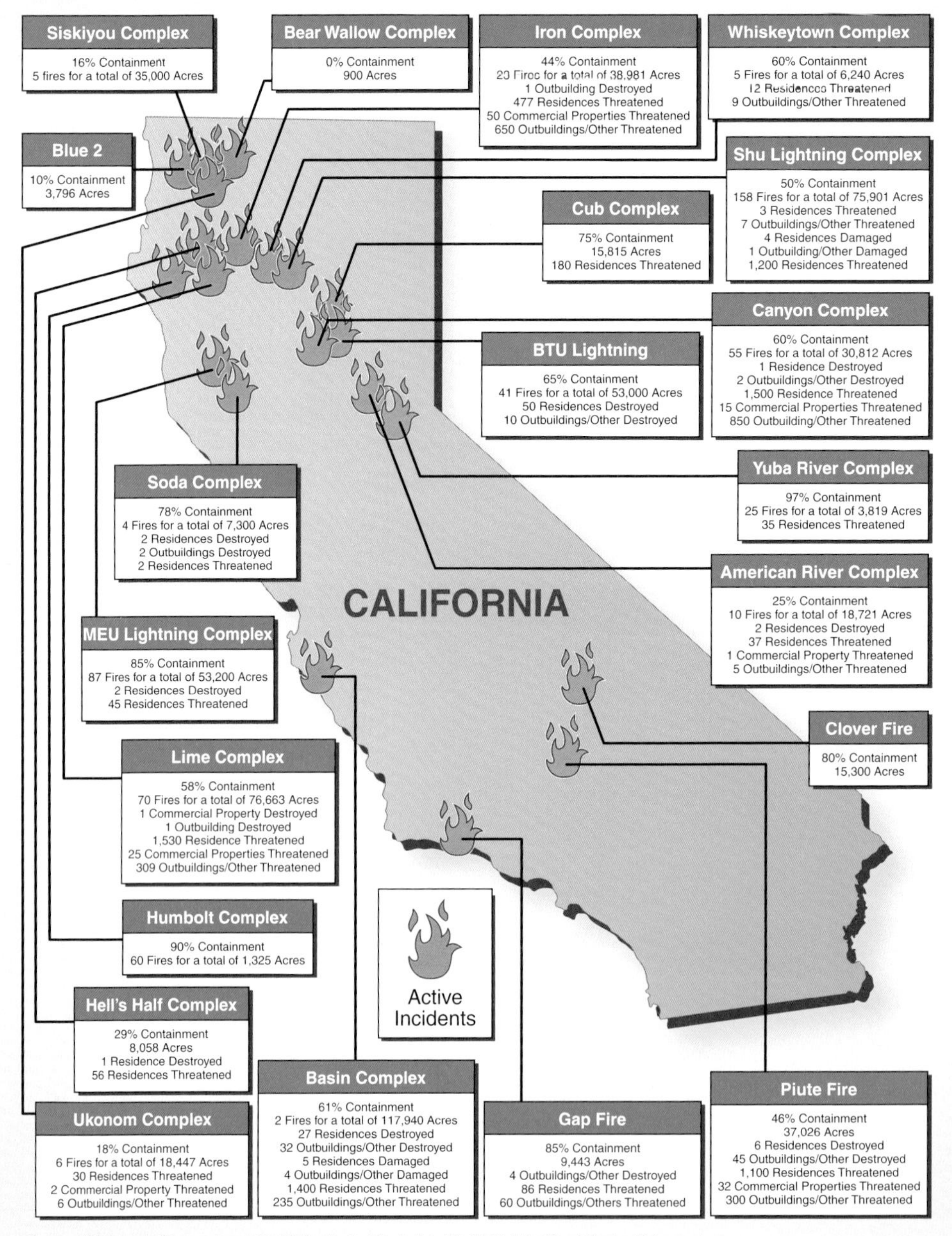

Source: "Pictures in Economics n' Stuff," The Daily Chart, July 13, 2008. http://thedailychart.blogspot.com.

Early Warning Tsunami System

The DART (Deep-Ocean Assessment and Reporting of Tsunamis) system, placed in the Indian Ocean in 2006, has the potential to save thousands of lives by warning of potential tidal waves. First (1), a recorder at the bottom of the ocean measures water pressure every 15 minutes. Next (2), a buoy measures surface conditions and sends this information plus data from the seabed to a satellite. Finally (3), the satellite sends the wave data to alert tsunami warning centers. This way, scientists can look for any unusual changes in sea level and chart the ocean's tidal patterns over time.

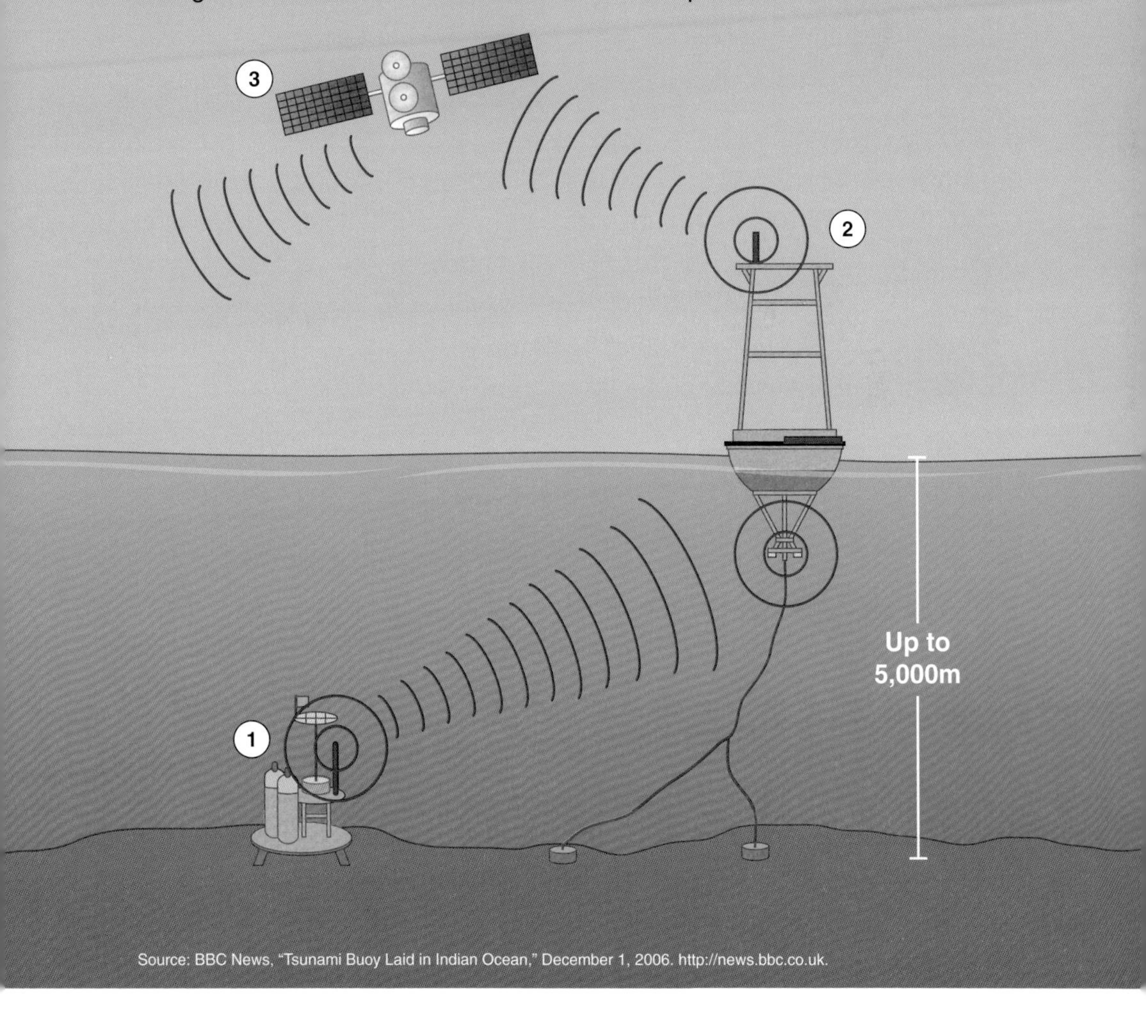

Source: BBC News, "Tsunami Buoy Laid in Indian Ocean," December 1, 2006. http://news.bbc.co.uk.

- Microsoft Corporation has developed software to **track refugees** fleeing war or natural disasters.

Key People and Advocacy Groups

American Red Cross: The American Red Cross, founded by Clara Barton in 1881, is a humanitarian organization led by volunteers, providing relief to victims of disasters. The American Red Cross also helps to prevent, prepare for, and respond to emergencies.

Catholic Charities: One of the largest social service networks in the nation, Catholic Charities provides aid and financial support to people in need, regardless of their religious, social, or economic backgrounds.

Center of Excellence in Disaster Management and Humanitarian Assistance (COE): A partnership of the United States Pacific Command, the Pacific Health Services Support Area of Tripler Army Medical Center, and the University of Hawaii, COE provides education, training, and research regarding international disaster management and humanitarian assistance and focuses on the Asia-Pacific region.

Joseph Cirincione: Former vice president for national security and international policy at the Center for American Progress in Washington, D.C., Cirincione is one of the world's leading experts on nuclear terrorism and proliferation.

Federal Emergency Management Agency (FEMA): This agency falls under the Department of Homeland Security of the United States and is charged with coordinating responses to natural and man-made disasters that overwhelm state and local resources.

Christophe Fournier: In 2006 Fournier became president of the humanitarian aid organization Médecins sans Frontières (MSF). Before then he worked as a doctor for MSF in Uganda, Honduras, Chile, and Burundi. In 2000 Fournier became MSF's operations manager, working in the United States and directing field programs in Haiti, Guatemala, Sudan, and Cambodia, among others.

Janet Napolitano: Former Arizona governor Napolitano is the current secretary of Homeland Security in the United States. Her early priorities included advanced security technology to protect against terrorist attacks.

Operation Blessing: An international relief and humanitarian organization providing food, clothing, shelter, medical care, and other basic necessities of life, Operation Blessing was established in 1978 and is governed by a national board of directors that includes founder M.G. Robertson.

Samaritan's Purse: This worldwide, nondenominational Christian organization provides assistance to victims of war, poverty, natural disasters, disease, and hunger. The organization was established in 1970 by Franklin Graham, the eldest son of televangelist Billy Graham.

UNICEF (Unite for Children): UNICEF is an advocate for the rights of children throughout the world. Its goals include reducing the number of children living in poverty and protecting them from the world's deadliest—and often most preventable—diseases.

United Nations Refugee Agency (UNHCR): The Office of the United Nations High Commissioner for Refugees is mandated to lead and coordinate international action to protect refugees and resolve refugee problems worldwide. Its primary purpose is to safeguard the rights and well-being of refugees.

World Health Organization: This United Nations agency, founded in 1948, coordinates international responses to outbreaks of infectious diseases and seeks to promote the health of people throughout the world.

Chronology

1863
The International Red Cross is founded by representatives of 16 nations meeting in Geneva, Switzerland.

1865
Direct federal participation in disaster relief begins when the U.S. Army Corps of Engineers helps freed blacks survive flooding along the Mississippi River.

1889
The South Fork Dam fails near Johnstown, Pennsylvania, flooding the town with 20 million tons of water and killing 2,200 people. Nurse and American Red Cross president Clara Barton arrives and stays for more than 5 months to aid victims of the disaster.

1906
A brief but powerful early morning earthquake strikes San Francisco, California, in April; the quake sparks 3 days of fires that destroy 490 city blocks, kills between 450 and 700 people, and leaves 250,000 survivors homeless.

1971
The UN Disaster Relief Office is founded to mobilize and coordinate relief activities in times of disaster.

1979
FEMA becomes the nation's primary federal agency responsible for disaster relief and assistance.

1984
Workers at a Union Carbide chemical plant in Bhopal, India, accidentally release 42 tons (38 metric tons) of toxic gas, exposing 500,000 to the deadly substance and immediately killing 2,259 people. Afterward, investigators found safety regulations at the plant to be lax.

1986
In what would become known as the worst nuclear power plant disaster in history, Ukraine's Chernobyl plant accidently releases high doses of radioactive material into the atmosphere. Workers in protective gear spend months collecting and containing the radioactive debris.

1989
After an oil tanker, the *Exxon Valdez*, spills 10.8 million gallons (40.9 million L) of crude oil into Alaska's Prince William Sound, an enormous cleanup effort begins to protect the wildlife of the area.

1990
The humanitarian aid organization Doctors Without Borders opens operations in the United States.

1900 1970 1980

1990

1991
A Bangladesh cyclone kills 130,000 people and leaves 9 million more homeless.

1992
Hurricane Andrew in South Florida costs the United States an estimated 26.5 billion dollars in damages; 1,500 National Guard troops are enlisted to prevent looting in the aftermath of the storm.

1993
War begins in Bosnia. Humanitarian groups work to feed and shelter thousands of refugees.

1998
Two major earthquakes in Afghanistan leave 10,000 dead and 45,000 homeless. The country's mountainous terrain makes relief work particularly challenging.

1999
Médecins sans Frontières wins the Nobel Peace Prize for its humanitarian work on several continents.

2000

2000
A December cyclone in Sri Lanka leaves 500,000 people homeless. It takes the government two days to activate rescue efforts; the media criticize officials for a slow response.

2001
On September 11, terrorists attack New York City and Washington, D.C. While attempting to save victims in the burning Twin Towers, 411 emergency workers die when the buildings collapse.

2002
Torrential rains across Europe cause severe flooding and kill dozens of people. Later, officials gather to discuss how to prevent such catastrophes in the future.

2005
Hurricane Katrina devastates Louisiana and Mississippi, killing more than 1,836 people and causing $81.2 billion in damage, including the destruction of approximately 275,000 homes.

2006
Seventeen aid workers are shot and killed by government troops in Sri Lanka.

2008
In November terrorists target hotels in Mumbai, India, killing 173 and wounding over 300. India's armed forces are faulted for their handling of the tragedy.

2009
In February Australian bushfires kill at least 200 people; firefighters battle the flames in an area more than twice the size of London.

Related Organizations

Action by Churches Together (ACT)

Ecumenical Centre
route de Ferney 150
PO Box 2100
CH-1211 Geneva 2
Switzerland
Web site: www.act-intl.org

ACT offers assistance to people caught in natural and environmental disasters, as well as in emergencies caused by war and civil conflict. Striving to reach communities in crises across front lines; national borders; and other ethnic, political or religious divides, ACT provides assistance irrespective of race, gender, belief, nationality, ethnic origin, or political persuasion.

Amnesty International

1 Easton St.
London WC1X 0DW, UK
phone: (+44 20) 7413 5500 • fax: (+44 20) 7956 1157
Web site: www.amnesty.org

Amnesty International is a worldwide movement of people who campaign for internationally recognized human rights for all. It has more than 2.2 million members and subscribers in more than 150 countries and regions and coordinates this support to act for justice on a wide range of issues.

Children's Disaster Services (CDS)

1451 Dundee Ave.
Elgin, IL 60120
phone: (800) 323-8039 or (847) 742-5100 • fax: (847) 742-6103
e-mail: cobweb@brethren.org • Web site: www.brethren.org

Children's Disaster Services trains, certifies, and mobilizes volunteers to provide crisis intervention to children suffering from natural or man-made disasters within the United States. CDS volunteers provide a safe and friendly environment to give children the freedom to act like children. In the child care center, children are given individualized attention while being engaged in therapeutic play activities designed to relieve stress and calm fears.

Doctors Without Borders

333 Seventh Ave., 2nd Floor
New York, NY 10001-5004
phone: (212) 679-6800 • fax: (212) 679-7016
e-mail: doctorswithoutborders.org

Doctors Without Borders is an international humanitarian aid organization that provides emergency medical assistance to populations in danger in more than 70 countries.

Federal Emergency Management Agency (FEMA)

500 C St. SW
Washington, DC 20472
Disaster Assistance: (800) 621-FEMA • TTY: (800) 462-7585
Web site: www.fema.gov

The primary mission of the Federal Emergency Management Agency is to reduce the loss of life and property and protect the United States from all hazards, including natural disasters, acts of terrorism, and other man-made disasters, by leading and supporting the nation in a risk-based, comprehensive emergency management system of preparedness, protection, response, recovery, and mitigation.

Feeding America

35 E. Wacker Dr., Suite 2000
Chicago, IL 60601
phone: (800) 771-2303 • fax: (312) 263-2656
Web site: feedingamerica.org

Feeding America is the nation's leading domestic hunger-relief charity. The Feeding America network supports approximately 63,000 local charitable agencies that distribute food directly to Americans in need. Those agencies operate more than 70,000 programs, including food pantries, soup kitchens, emergency shelters, and after-school programs.

Human Rights Watch

350 Fifth Ave., 34th Floor
New York, NY 10118-3299

phone: (212) 290-4700 • fax: (212) 736-1300
e-mail: hrwnyc@hrw.org • Web site: www.hrw.org

Human Rights Watch is, according to its Web site, "one of the world's leading independent organizations dedicated to defending and protecting human rights." It works to focus international attention when and where human rights are violated, and gives voice to oppressed peoples, and hold oppressors accountable for their crimes.

International Federation of Red Cross and Red Crescent Societies

PO Box 372
CH-1211 Geneva 19
Switzerland
phone: (+41 22) 730 42 22 • fax: (+41 22) 733 03 95
Web site: www.ifrc.org

The International Federation of Red Cross and Red Crescent Societies is the world's largest humanitarian organization, providing assistance without discrimination as to nationality, race, religious beliefs, class, or political opinions. The federation carries out relief operations to assist victims of disasters and combines this with development work to strengthen the capacities of its member national societies. The federation's work focuses on four core areas: promoting humanitarian values, disaster response, disaster preparedness, and health and community care.

Oxfam

226 Causeway St., 5th Floor
Boston, MA 02114-2206
phone: (617) 482-1211 • toll free: (800) 776-9326
fax: (617) 728-2594
e-mail: info@oxfamamerica.org • Web site: www.oxfam.org

Oxfam International is a confederation of 13 like-minded organizations working together and with partners and allies around the world to bring about lasting change. Working with more than 3,000 local partner organizations, Oxfam works with people living in poverty who strive to exercise their human rights, assert their dignity as full citizens, and take control of their lives. Among other things, Oxfam delivers immediate life-saving assistance to people affected by natural disasters or conflict and helps to build their resilience to future disasters.

Salvation Army

615 Slaters Ln.
PO Box 269
Alexandria, VA 22313
www.salvationarmyusa.org

This organization provides emergency assistance, including mass and mobile feeding, temporary shelter, counseling, missing person services, medical assistance, and distribution of donated goods, including food, clothing, and household items. It also provides referrals to government and private agencies for special services.

United Methodist Committee on Relief (UMCOR)

475 Riverside Dr., Room 330
New York, NY 10115
phone: (800) 554-8583
e-mail: umcor@gbgm-umc.org • Web site: www.umcor.org

UMCOR responds to natural or civil disasters that are interruptions of such magnitude that they overwhelm a community's ability to recover on its own. UMCOR's work on behalf of the United Methodist Church is global and includes countries in Africa, Asia, Central and South America, and the Caribbean. UMCOR serves in long-term disaster recovery in the United States.

World Health Organization (WHO)

Ave. Appia 20
1211 Geneva 27
Switzerland
phone: (+41 22) 791 21 11 • fax: (+41 22) 791 31 11
Web site: www.who.int

WHO is the directing and coordinating authority for health within the United Nations system. It is responsible for providing leadership on global health matters, shaping the health research agenda, setting norms and standards, articulating evidence-based policy options, providing technical support to countries, and monitoring and assessing health trends.

For Further Research

Books

Joseph Cirincione, *Bomb Scare: The History and Future of Nuclear Weapons*. New York: Columbia University Press, 2007.

Stephen Flynn, *The Edge of Disaster: Rebuilding a Resilient Nation*. New York: Random House, 2007.

Bruce Hoffman, *Inside Terrorism*. New York: Columbia University Press, 2006.

Jed Horne, *Breach of Faith: Hurricane Katrina and the Near Death of a Great American City*. New York: Random House, 2008.

Erich Krauss, *Wave of Destruction: The Stories of Four Families and History's Deadliest Tsunami*. Emmans, PA: Rodale, 2005.

Sarah Kenyon Lischer, *Dangerous Sanctuaries: Refugee Camps, Civil War, and the Dilemmas of Humanitarian Aid*. Ithaca, NY: Cornell University Press, 2006.

James F. Miskel, *Disaster Response and Homeland Security: What Works, What Doesn't*. Palo Alto, CA: Stanford University Press, 2008.

Steven VanOrden, *Africa: Stranger than Fiction: Memoirs of a Humanitarian Aid Worker*. Bloomington, IN: iUniverse, 2008.

Simon Winchester, *A Crack in the Edge of the World: America and the Great California Earthquake of 1906*. New York: HarperPerennial, 2006.

Periodicals

Associated Press, "Mobile Conference Puts Focus on Using Technology for Disaster Response, Economic Improvements," *International Herald Tribune*, February 14, 2008. www.iht.com.

Christian Science Monitor, "In Burma (Myanmar), How Many Cyclone Orphans?" June 9, 2008.

Shaila Dewan, "Many Children Lack Stability Long After Storm," *New York Times*, December 4, 2008. www.nytimes.com.

Jacob Goodwin, "How Do First Responders React When All Hell Breaks Loose?" *Government Security News*, August 11, 2008. www.gsnmagazine.com.

Spencer S. Hsu and Ann Scott Tyson, "Pentagon to Detail Troops to Bolster Domestic Security," *Washington Post*, December 1, 2008.

Robert Krier, "Ready for the Big One: 5 Million Take Part in Earthquake Drill," *San Diego Union-Tribune*, November 14, 2008.

Sara Miller Llana, "After Flood, Long-Term Test for Mexico," *Christian Science Monitor*, November 16, 2007.

Karen Matthews, "Survey Finds Holes in U.S. Disaster Preparedness," *USA Today*, September 12, 2008. www.usatoday.com.

Walter Pincus, "Pnel Cites 'Tipping Point' on Nuclear Proliferation," *Washington Post*, December 16, 2008.

James Pinkerton, "Hurricane Ike: The Aftermath," *Houston Chronicle*, October 4, 2008.

David R. Sands, "Panel Warns of Risk of an Attack by 2013," *Washington Times*, December 3, 2008.

Somini Sengupta, "At Least 100 Dead in India Terror Attacks," *New York Times*, November 26, 2008. www.nytimes.com.

Mark Waller, "FEMA Has Worked Hard to Rebuild Agency," *New Orleans Times-Picayune*, November 4, 2008.

Anthony R. Wood, "Hurricanes Prove Costly in 2008," *Philadelphia Inquirer*, November 27, 2008.

Robert F. Worth, "Lack of Preparedness Comes Brutally to Light," *New York Times*, December 4, 2008.

Internet Sources

Christi Harlan, "Kentucky Governor Thanks Red Cross for Help During State's Biggest Natural Disaster," *American Red Cross*, February 4, 2009. www.redcross.org/portal/site/en/menuitem.1a019a978f421296e81ec89e43181aa0/?vgnextoid=87fcdd081524f110VgnVCM10000089f0870aRCRD.

Médecins san Frontières, "MSF Responding to Worst Cholera Outbreak in Zimbabwe in Years," December 15, 2008. www.msf.org/msfinternational/invoke.cfm?objectid=3A06C714-15C5-F00A25B61038A975AFAB&component=toolkit.article&method=full_html.

Oxfam International, "DR Congo: Groups Fear for Civilian Safety," February 6, 2009. www.oxfam.org/en/pressroom/pressrelease/2009-02-06/dr-congo-groups-fear-for-civilian-safety.

Gary Thomas, "Analysts Monitor Political Impact of Disaster Response in China, Burma," *Voice of America*, May 26, 2008. www.voanews.com/english/archive/2008-05/2008-05-26-voa18.cfm?CFID=75562729&CFTOKEN=47639596.

UNICEF Press Centre, "Increasing Number of Children Injured in Fighting in Sri Lanka," January 30, 2009. www.unicef.org/media/media_47673.html.

Mike Von Fremd et al., "State of Emergency in California as Blaze Rages," ABC News, November 14, 2008. http://abcnews.go.com/Travel/story?id=6253130&page=1.

Graham Wood, "An Aid Worker's Life in Kabul," Reuters Foundation, August 14, 2007. www.alertnet.org/db/blogs/44345/2007/07/14-142318-1.htm.

Source Notes

Overview

1. Quoted in Douglas Brinkley, *The Great Deluge*. New York: HarperCollins, 2006, p. 344.
2. Quoted in Somini Sengupta and Keith Bradsher, "India Faces Reckoning as Terror Toll Eclipses 170," *New York Times*, November 29, 2008. www.nytimes.com.
3. Quoted in PJ Heller, "Faith Organizations' Ready Response," *Disaster News Network*, September 1, 2008. www.disasternews.net.
4. Quoted in "Annan Says NGOs Essential to Work of UN," *Agence France Presse*, August 29, 2000. www.globalpolicy.org.
5. Tim Russell, "The Humanitarian Relief Supply Chain: Analysis of the 2004 Southeast Asia Earthquake and Tsunami," MIT Center for Transportation and Logistics, 2005, p. 1.
6. Quoted in Jonathan Harr, "Lives of the Saints," *New Yorker*, January 5, 2009, p. 52.
7. Quoted in "MSF Responding to Worst Cholera Outbreak in Zimbabwe in Years," MSF Article, December 15, 2008. www.msf.org.
8. Tom Halloran, interview, January 6, 2009, with David Robson. Wilmington, DE.
9. Quoted in Deborah Tate, "US Takes Steps to Prepare for Potential Nuclear Attack," Voice of America, June 26, 2008. www.voanews.com.
10. Quoted in *Panorama*, "Coping with Biological Threat." http://news.bbc.co.uk.
11. Quoted in Voice of America, "Analysts Monitor Political Impact of Disaster Response in China, Burma," May 26, 2008. www.voanews.com.
12. Quoted in Reuters Africa, "Zimbabwe Cholera Deaths Not Slowing; 1,600 Dead," December 30, 2008. http://africa.reuters.com.
13. Quoted in Jonathan Harr, "Lives of the Saints," *New Yorker*, January 5, 2009, p. 55.
14. Stephen Flynn, "Flynn: U.S. Not Prepared for the Next 'Big One,'" CNN, February 27, 2007. www.cnn.com.

How Does the World Respond to Disasters?

15. Quoted in Dan Bortolotti, *Hope in Hell*. Buffalo, NY: Firefly, 2004, pp. 201, 202.
16. Russell, "The Humanitarian Relief Supply Chain," p. 4.
17. Damon P. Coppola, *Introduction to International Disaster Management*. St. Louis, MO: Butterworth-Heinemann, 2006, p. 527.
18. Jan Egeland, *Billions of Lives: An Eyewitness Report from the Frontlines of Humanity*. New York: Simon and Schuster, 2008, p. 106.
19. Quoted in BBC News, "Corruption 'Mars Iraq Rebuilding,'" July 30, 2007. http://news.bbc.co.uk.
20. Quoted in CNN, "Afghan Aid Workers Raise Security After Killing," October 21, 2008. www.cnn.com.
21. Quoted in *The Independent*, "UN Suspends Gaza Aid over Fears for Staff," January 8, 2009. www.independent.co.uk.
22. Quoted in Nicole Johnston, "When Fragile Lives Are Ripped Apart by Disaster," *Mail & Guardian Online* (South Africa), March 9, 2007. www.mg.co.za.

23. Quoted in ICRC, "Democratic Republic of the Congo: Charlotte's Smile," August 12, 2008. www.icrc.org.

How Does the United States Respond to Natural Disasters?

24. Bruce Schreiner and Betsy Taylor, "Guard Call-up to Aid 600,000 in Kentucky Left Without Power," Associated Press/*Philadelphia Inquirer*, February 1, 2009, p. A16.
25. Quoted in Spencer S. Hsu, "Katrina Report Spreads Blame," *Washington Post*, February 12, 2006, p. A01.
26. Quoted in Manuel Roig-Franzia and Spencer Hsu, "Many Evacuated, but Thousands Still Waiting," *Washington Post*, September 4, 2005, p. A01.
27. Quoted in "Kelley Shannon, "Groups That Aided Ike Storm Victims Still Awaiting Pay," *Philadelphia Inquirer*, February 25, 2009, p. A4.
28. Benjamin Wisner, *At Risk: Natural Hazards, People's Vulnerability and Disasters*. New York: Routledge, 2003, p. 180.
29. Quoted in Randal A. Archibold and Kirk Johnson, "Anxiety Grows in West over Firefighting Efforts," *New York Times*, June 19, 2008. www.nytimes.com.
30. Quoted in American Red Cross, "How to Touch Lives," 2007. www.redcrossdallas.org.

Is the United States Prepared for Man-Made Disasters?

31. Quoted in Frank Sesno,"Could Smallpox Really Happen?" CNN, September 26, 2005. www.cnn.com.
32. Quoted in Ewen MacAskill, "Terrorists Could Mount Nuclear or Biological Attack Within 5 Years, Warns Congress Inquiry," *Guardian*, December 4, 2008. www.guardian.co.uk.
33. Quoted in Chris Jablonski, "The Impact of Nuclear Attacks on U.S. Cities," ZDNet, March 24, 2007. http://blogs.zdnet.com.
34. Graham Allison, *Nuclear Terrorism: The Ultimate Preventable Catastrophe.*" New York: Times Books, 2004, p. 198.
35. Quoted in Matthew L. Wald, "Faulty Design Led to Minnesota Bridge Collapse, Inquiry Finds," *New York Times*, January 15, 2008. www.nytimes.com.
36. Mortimer B. Zuckerman, "Obama Economic Stimulus Offers Too Much Waste, Not Enough Job Creation," *U.S. News & World Report*, February 16, 2009. www.usnews.com.
37. Quoted in David Grann, "City of Water," *New Yorker*, September 1, 2003, p. 91.
38. Quoted in CNN, "Hudson River Plane Rescuers Still in Disbelief a Day Later," January 16, 2009. www.cnn.com.

How Can Disaster Response Be Improved?

39. Quoted in Mark Tran and Elizabeth Stewart, "China Earthquake Kills Thousands," *Guardian*, May 12, 2008. www.guardian.co.uk.
40. Quoted in China, "China Prepares to Improve Disaster Response and Relief Capability," October 11, 2006. www.china.org.cn.
41. Russell, "The Humanitarian Relief Supply Chain," p. 7.
42. Quoted in Julian Borger, "Lives Lost Through Lack of Leadership in UN Response to Humanitarian Crises, Britain Warns," *Guardian*, October 7, 2008. www.guardian.co.uk.
43. Quoted in Department of Homeland Security, press release, "Secretaries Gutierrez, Chertoff Announce Nearly $1 Billion in First Responder Communications Grants Funds to Help

Fire Fighters, Police and Other First Responders Communicate During a Disaster," July 18, 2007. www.dhs.gov.
44. Quoted in United Nations Foundation, "Mobile Activism: Make Text, Not War," January 2008. www.unfoundation.org.
45. Quoted in Chris Strohm, "White House: Agencies Remain Unprepared for Disasters," *Government Executive*, February 23, 2006. www.govexec.com.
46. Lawrence P. Farrell Jr., "Preparation Is Key to Disaster Response," *National Defense*, November 2005. www.nationaldefensemagazine.org.
47. Quoted in Tara Parker-Hope, "Learning to Be Your Own Best Defense in a Disaster, *New York Times*, August 5, 2008. www.nytimes.com.
48. Quoted in Robert Krier, "Ready for the Big One," *San Diego Union-Tribune*, November 14, 2008, p. B-1.
49. Quoted in *Government Technology*, "Health Organizations Release Recommendations for Improved Disaster Response," July 20, 2007. www.govtech.com.
50. Quoted in Heather B. Hayes, "Comcare, HIMSS Form Board to Promote Emergency Integration," *Government Health IT*, February 21, 2008. http://govhealthit.com.
51. Victor Renuart, "Disaster Response Needs More Collaboration," *Middle East Times*, December 18, 2008. www.metimes.com.
52. Quoted in UCSD Jacobs School of Engineering, "UCSD Researchers Test Wireless," May 16, 2005. www.jacobsschool.ucsd.edu.

List of Illustrations

Index

About the Author

David Robson is the recipient of two playwriting fellowships from the Delaware Division of the Arts. His plays have been performed across the country. He is also the author of several books for young adults, including *Auschwitz, The Black Arts Movement*, and *The Kennedy Assassination*. Robson holds an MFA from Goddard College, an MS from Saint Joseph's University, and a BA from Temple University. He lives with his family in Wilmington, Delaware.